WORKBOOKS IN CHEMISTRY

SERIES EDITOR

STEPHEN K. SCOTT

WORKBOOKS IN CHEMISTRY

Beginning mathematics for chemistry
Stephen K. Scott

Beginning calculations in physical chemistry
Barry R. Johnson and Stephen K. Scott

Beginning organic chemistry 1
Graham L. Patrick

Beginning organic chemistry 2
Graham L. Patrick

WORKBOOKS IN CHEMISTRY

Beginning Organic Chemistry 1

GRAHAM L. PATRICK

Department of Chemistry
Paisley University

OXFORD NEW YORK TOKYO
OXFORD UNIVERSITY PRESS
1997

Oxford University Press, Great Clarendon Street, Oxford OX2 6DP

Oxford New York
Athens Auckland Bangkok Bogota Bombay Buenos Aires
Calcutta Cape Town Dar es Salaam Delhi Florence Hong Kong
Istanbul Karachi Kuala Lumpur Madras Madrid Melbourne
Mexico City Nairobi Paris Singapore Taipei Tokyo Toronto
and associated companies in
Berlin Ibadan

Oxford is a trade mark of Oxford University Press

Published in the United States
by Oxford University Press Inc., New York

A catalogue record for this book is available from the British Library

Library of Congress Cataloging in Publication Data
(Data applied for)

ISBN 0 19 855935 6

Typeset by the author
Printed in Great Britain by
Progressive Printing UK Ltd
Leigh-on-Sea, Essex

Workbooks in Chemistry: Series Preface

The new *Workbooks in Chemistry Series* is designed to provide support to students in their learning in areas that cannot be covered in great detail in formal courses. The format allows individual, self-paced study. Students can also work in groups guided by tutors. Teaching staff can monitor progress as the students complete the exercises in the text. The Workbooks aim to support the more traditional teaching methods such as lectures. The format of the Workbooks has been evolved through experience and discussions with students over several years. Students benefit through the Examples and Exercises that provide practice and build confidence. University staff faced with increasing class sizes may find the Workbooks helpful in encouraging 'self-learning' and meeting the individual needs of their students more efficiently. The topics covered in the early Workbooks in the Series will concentrate on background support appropriate to the early years of a Chemistry degree, including mathematics, performing calculations, and basic concepts in organic chemistry. These should also be of interest to students who are taking chemistry courses as part of other degree schemes, such as biochemistry and environmental sciences. Later Workbooks will be designed to support material typically encountered in later years of a Chemistry course.

Contents

Introduction

The following self-learning text has been prepared to help you in your studies of organic chemistry. By working through the various sections, you will be led through some of the basics of organic chemistry. You will also get plenty of practice at problem solving and answering the sort of questions you might get asked in an examination. In fact, probably one of the greatest advantages of using this text is that it gives you model answers for all the problems and questions asked and lets you see what examiners expect of you when they set a question. Too often, students fail to do themselves justice in an exam because they have either not put down enough in their answer or have strayed from the point.

The text is split into various sections as follows:

Section 1	Drawing Structures
Section 2	Functional Groups
Section 3	Formulae and Molecular Weights
Section 4	Nomenclature—Alkanes
Section 5	Nomenclature—Functional Groups
Section 6	Primary, Secondary, Tertiary and Quaternary Nomenclature
Section 7	Constitutional Isomers—Alkanes
Section 8	Constitutional Isomers Containing Functional Groups
Section 9	Non-equivalent Carbons and Hydrogens 1
Section 10	Non-equivalent Carbons and Hydrogens 2
Section 11	Configurational Isomers—Alkenes
Section 12	Configurational Isomers—Optical Isomers
Section 13	Conformational Isomers
Section 14	Shape and Bonding

In each section, you will get a short bit of theory, then a question. An empty box will be provided for your answer and the correct answer follows on. To get the best out of this text, you should read your lecture notes and textbook on the subject you wish to study, then find the relevant section in the self-learning text. Try to answer the question before looking at the answer and use a pencil so that you can reuse the self-learning text at a later date. If you find it difficult to resist the temptation of 'sneaking a look' at the answer, you could always paste pieces of paper over the answers, which you can then fold back as and when you wish to look at the answer. Don't worry if you don't get the correct answers the first time you tackle a particular section. You would be exceptional if you did. The important thing is to try the questions and to come up with your own answers. Once you have identified your mistakes and you repeat the section at a later stage, you will make fewer errors. Your goal is to keep repeating the section until you don't make any errors at all and can answer the questions in the way shown. In this way, you should get all your mistakes and errors ironed out before the exam comes along.

Paisley
July 1996

Graham L. Patrick

SECTION 1

Drawing Structures

Drawing Structures

There are several ways of drawing organic molecules. You may already have written a molecule such as ethane by showing every bond in the structure. This takes a long time and it is much easier to miss out the C–H bonds, as shown below.

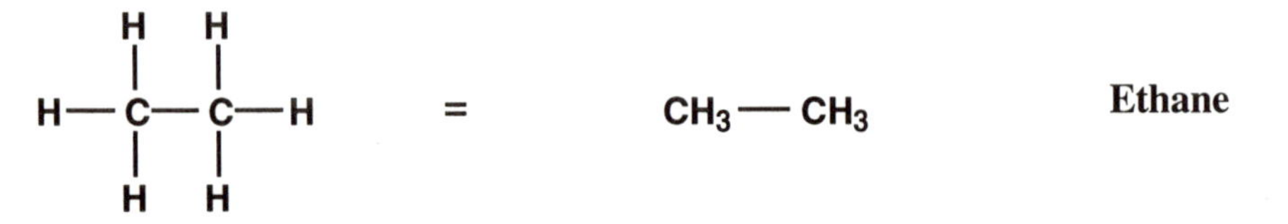

It is important that you can use this shorthand method of drawing molecules. Try and draw the following structures in this shorthand fashion.

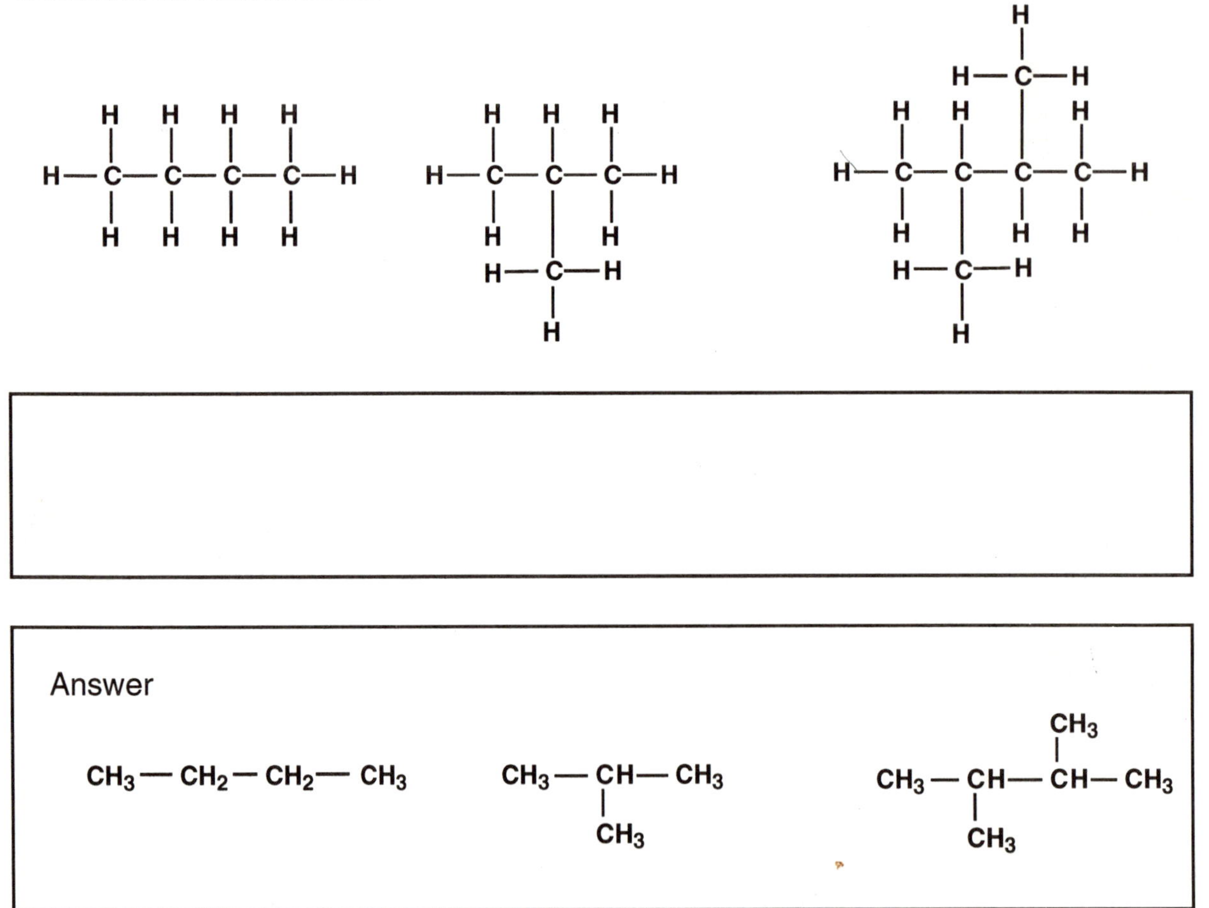

A further simplification is often used for more complicated molecules and for molecules with rings. The molecules are drawn without showing the carbon or hydrogen atoms—only the carbon to carbon bonds are shown. This is known as the carbon skeleton of the molecule. When you see such a drawing, it is understood that a carbon atom is present at every bond junction and that every carbon has enough hydrogens on it such that it has four bonds.

Example

What is the full structure of the following compound?

Answer

(1) Each bond junction has a carbon atom at it.
(2) Each carbon must have four bonds. Therefore, add hydrogens to make four bonds.

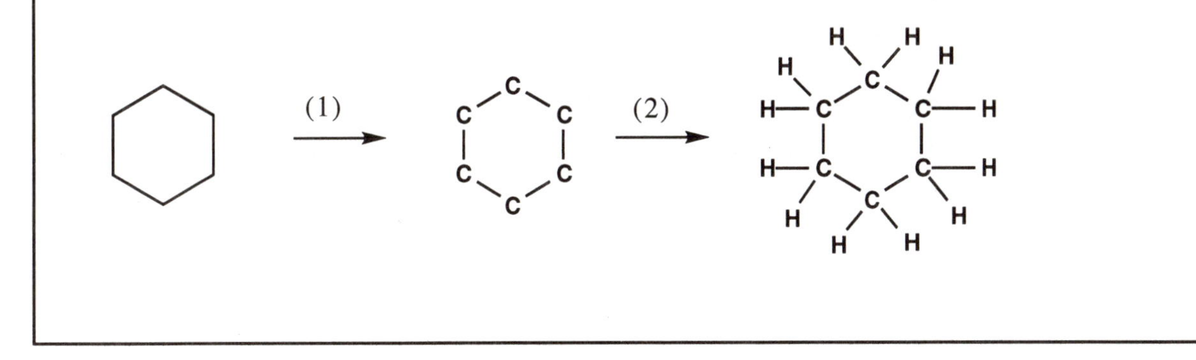

To give you practice at drawing simple carbon skeletons, simplify the following full structures.

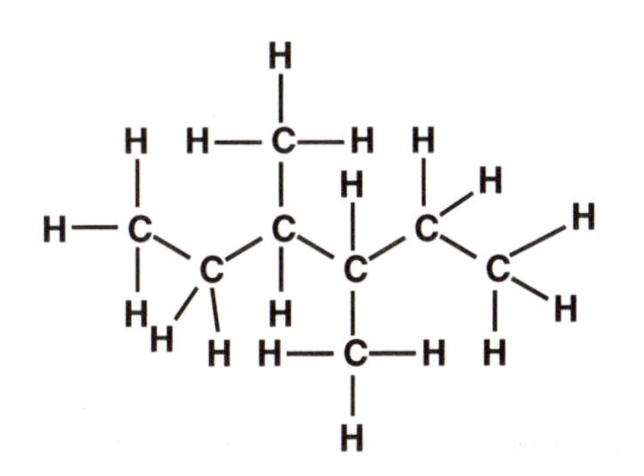

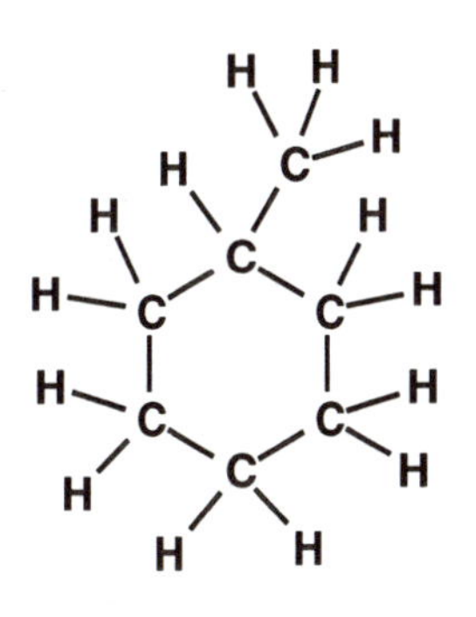

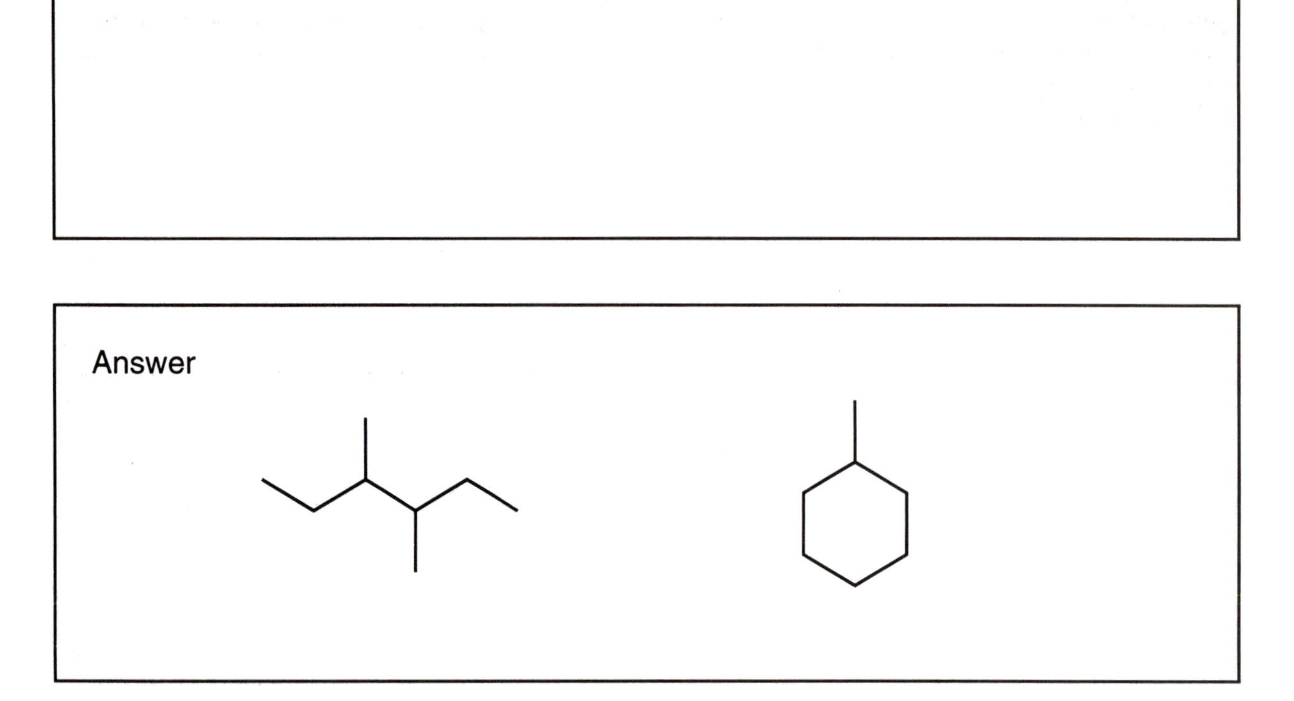

It should be obvious that these drawings are quicker to draw and are much clearer. Now simplify the following structures.

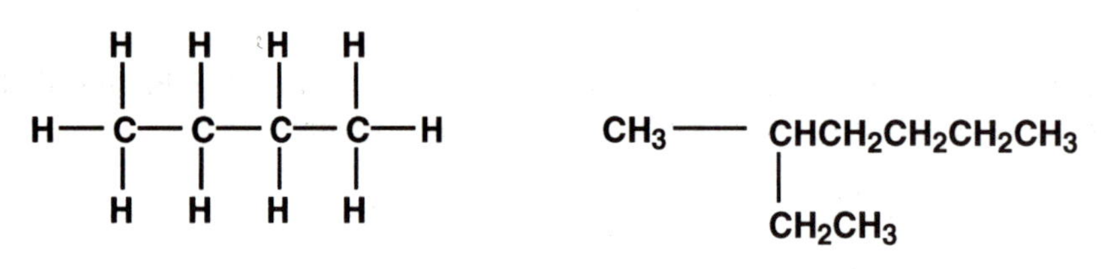

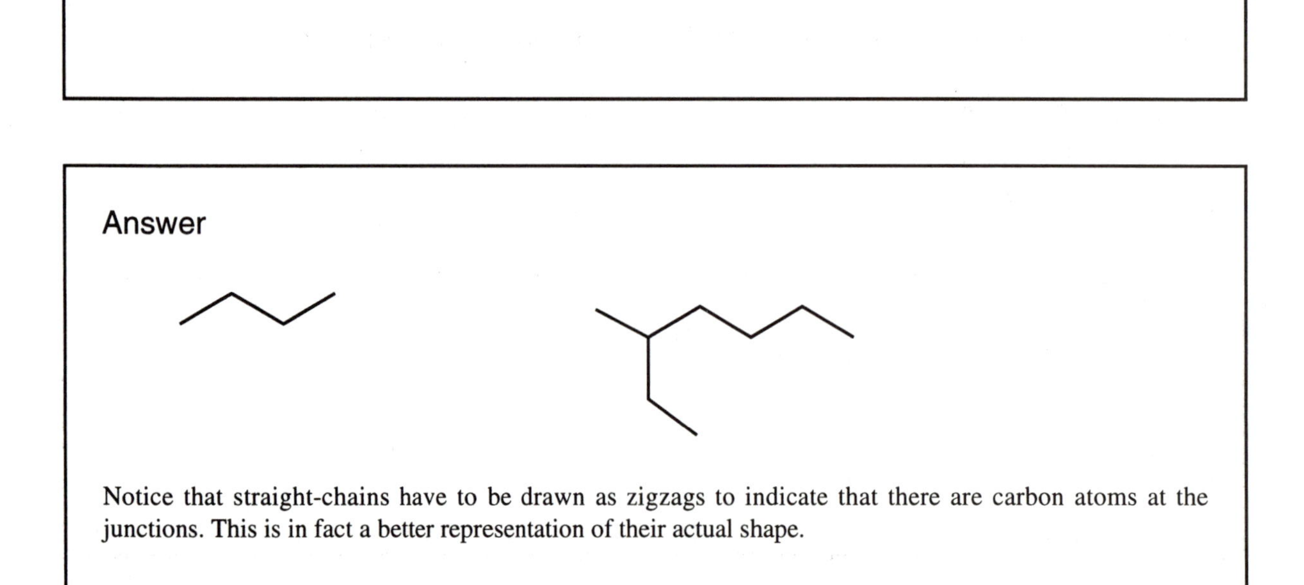

Notice that straight-chains have to be drawn as zigzags to indicate that there are carbon atoms at the junctions. This is in fact a better representation of their actual shape.

It should be clear to you by now that putting in all the carbons and hydrogens for a complicated molecule makes the drawing very confusing and also takes a long time to draw. It is much easier and more convenient to use the simple 'skeleton' drawings. However, it is important to remember that the carbons and hydrogens are still there, especially if you wish to find out the molecular formula and molecular weight of the compound. In the next couple of questions, you are asked for the molecular formula and molecular weights of different alkanes and cycloalkanes which have been written in the skeleton form.

What are the molecular formulae and weights of the following molecules?

Answer

In time you will be able to answer this type of question without adding atoms, but to begin with it is worth putting in the carbons and hydrogens to make sure you do not miss anything.

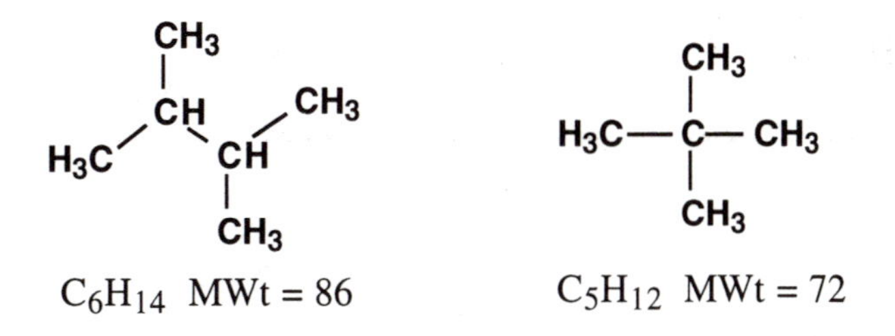

The following molecules are more complex. What are the molecular formulae and weights?

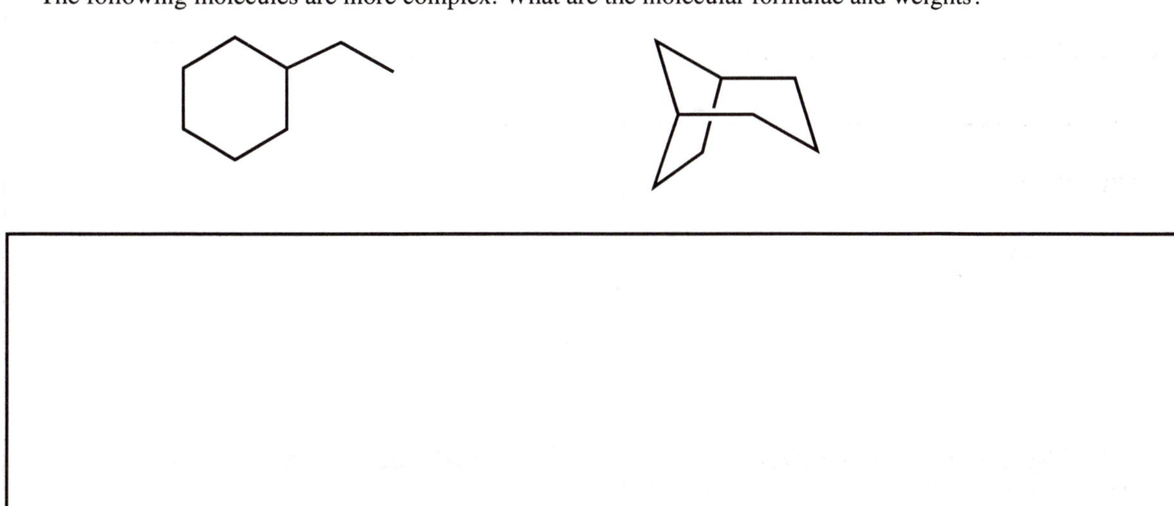

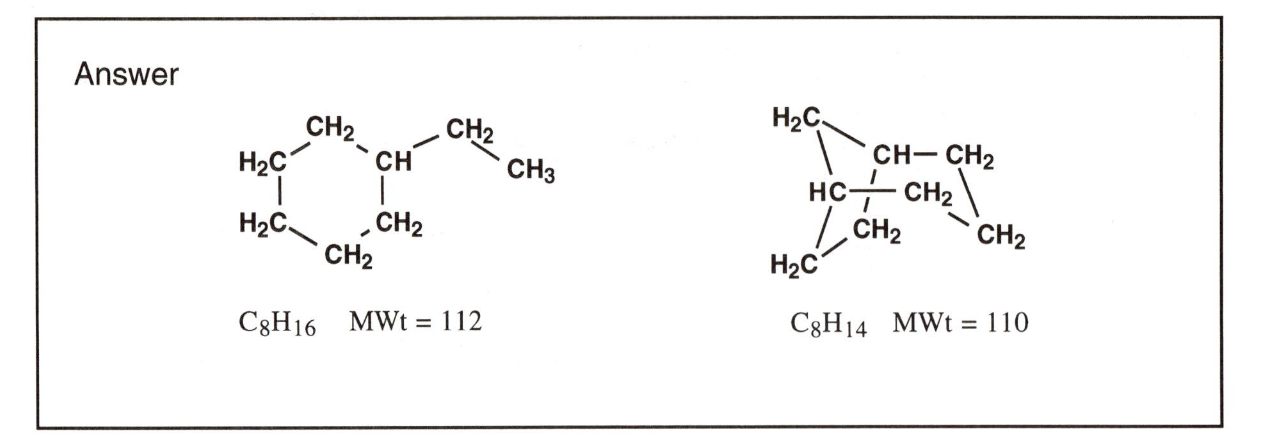

Often you will see both forms of structure shorthand used in the same molecule, but it is generally the case that ring structures are drawn in the simplest way possible.

Examples

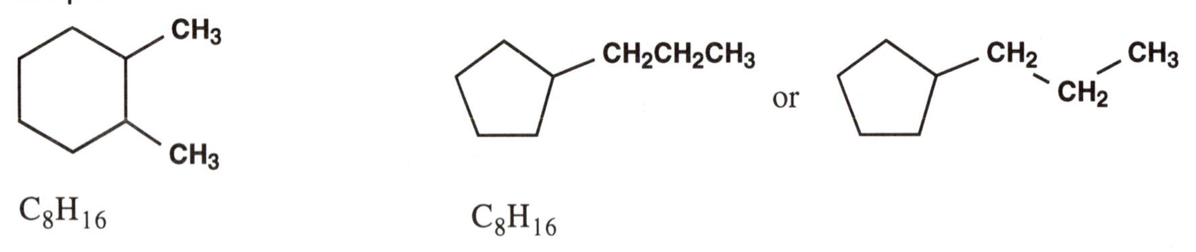

C_8H_{16} C_8H_{16}

It is worth checking that you can locate all the carbons and hydrogens in these structures.

Simplify the following molecule in the manner described above.

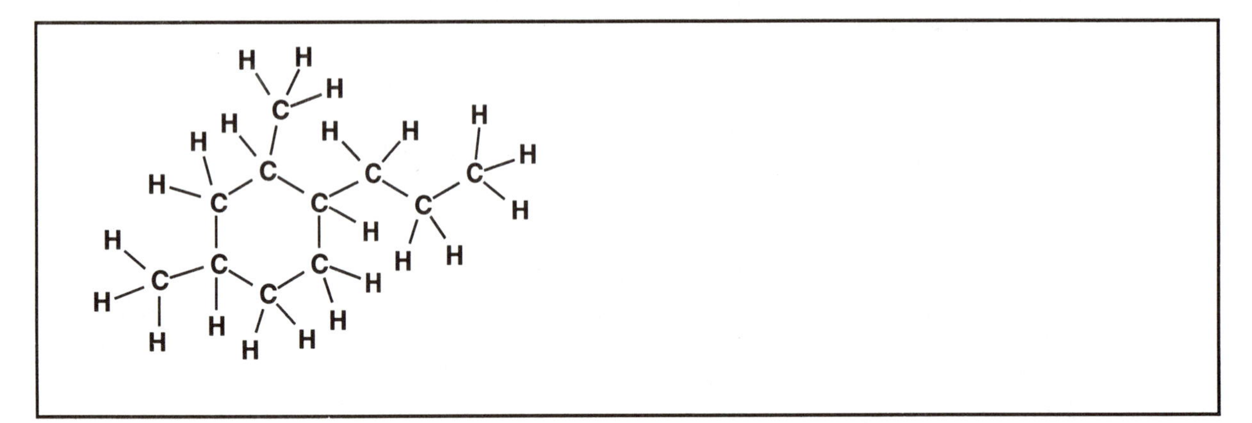

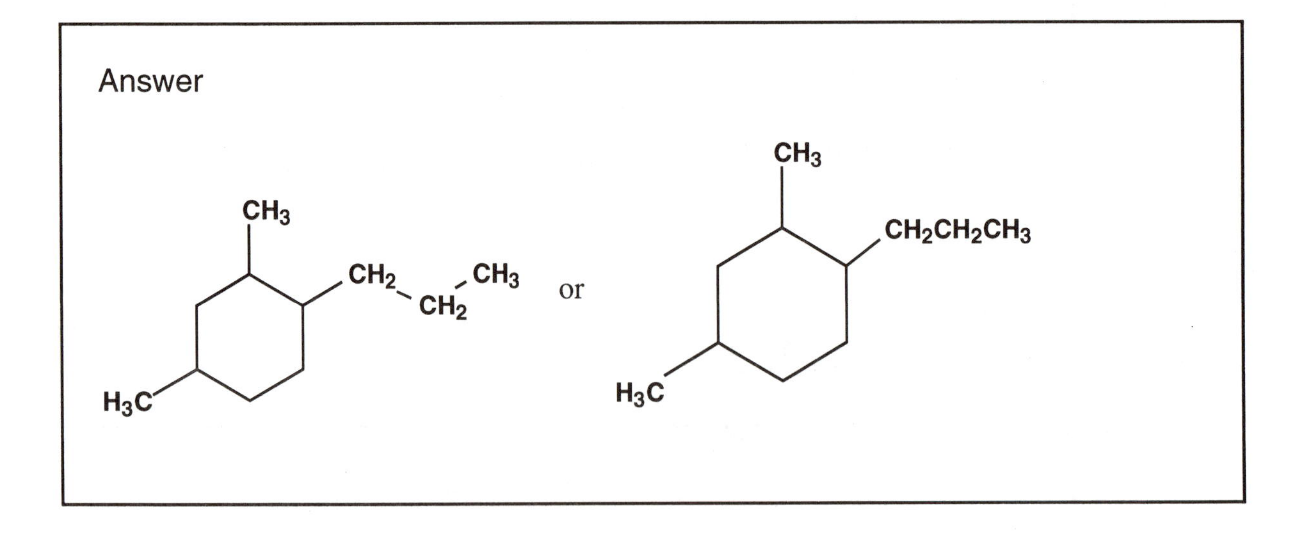

Notice that the carbon–carbon bonds in the alkyl chain have been missed out on the right-hand structure. This is an acceptable way of representing simple alkyl chains. Notice also how the CH_3 groups have been written. The structure on the left (below) is more correct than the structure on the right (below) since the bond shown is between the carbons, not the hydrogens.

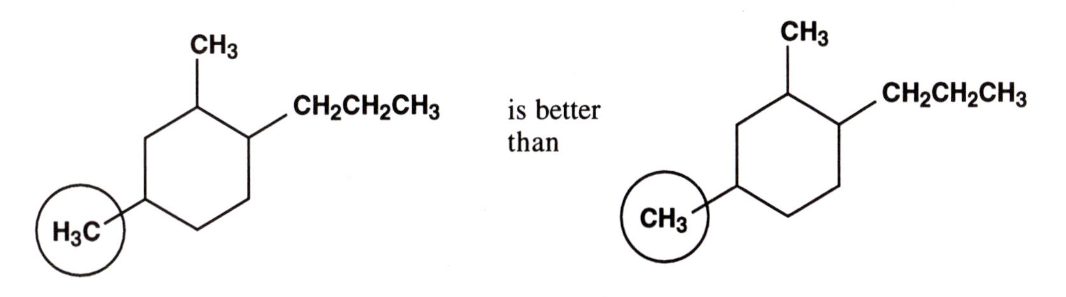

However, you would rarely be marked wrong for drawing a molecule as shown on the right.

Summary

- Organic molecules are normally drawn without showing the carbon–hydrogen bonds.
- The simplest method of drawing organic molecules is the skeletal form where only the carbon–carbon bonds are shown and where the carbon and hydrogen atoms are not shown. Each bond junction represents a carbon atom. The number of hydrogens present at each carbon will be the number needed for carbon to have four bonds.
- Rings are normally represented by the simple skeletal form.
- Alkyl chains can be represented by a simple skeletal form or as a group of atoms such as CH_2CH_3 etc.

SECTION 2

Functional Groups

Functional Groups

2.1 Identifying functional groups

There are a large number of reactions in organic chemistry and learning them all is no easy matter. However, we can make things a bit easier by appreciating that the vast majority of organic reactions take place at **functional groups**.

What is a functional group?

A functional group is a portion of an organic molecule, other than the normal hydrocarbon framework. For example, consider the following molecules, ethane and ethanoic acid.

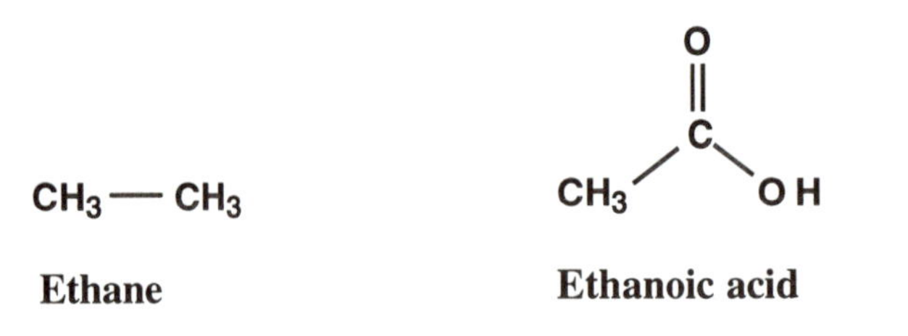

Ethane is a saturated hydrocarbon (an alkane) and has no functional group. Ethanoic acid, on the other hand, has a functional group known as a carboxylic acid outlined in the boxed area below.

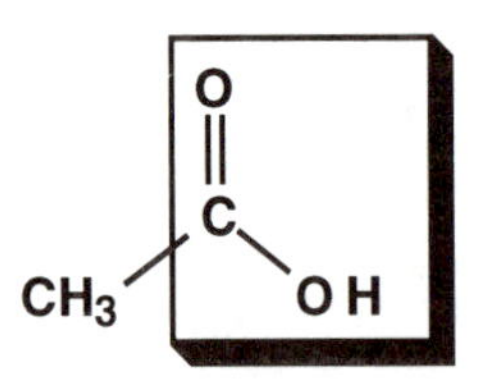

Carboxylic acid functional group

Whenever you see this arrangement of atoms again, you will know it represents a carboxylic acid functional group. The rest of the molecule is not part of the functional group and we can represent it by R as shown below. R stands for the hydrocarbon or alkyl part of the molecule.

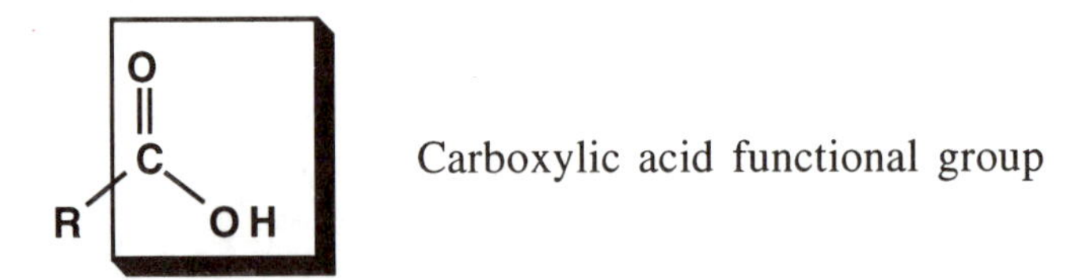

The next exercise gives you practice at identifying those molecules which contain functional groups and those which do not. Remember a functional group is any portion of the molecule which is not made up of carbon–carbon or carbon–hydrogen single bonds. Which of the following structures contain functional groups? Outline the functional groups which are present in the following structures.

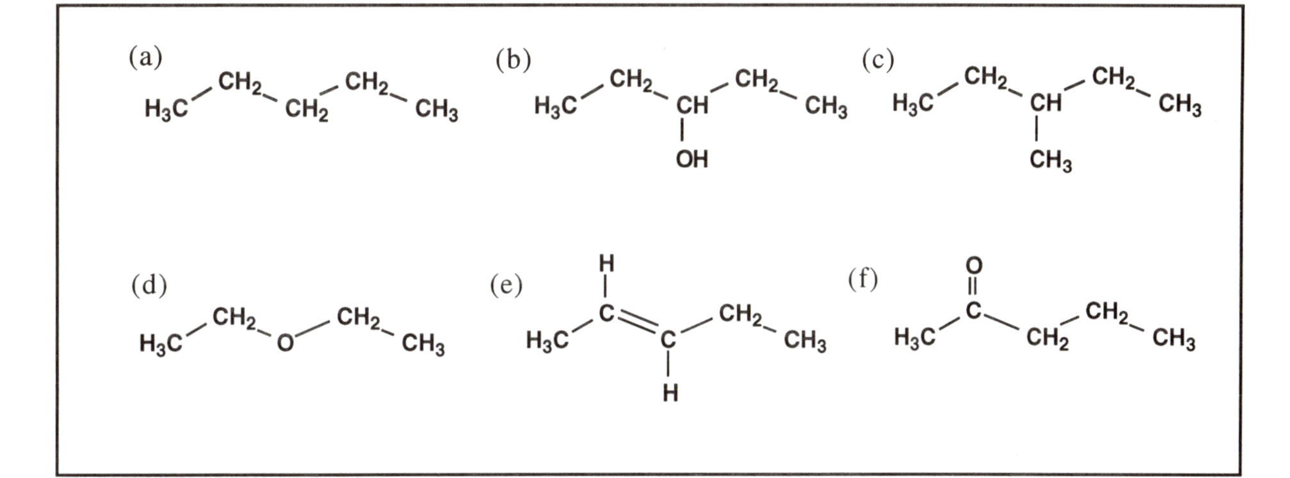

Answer

Compounds (b), (d), (e) and (f) all contain functional groups. The functional groups are outlined below.

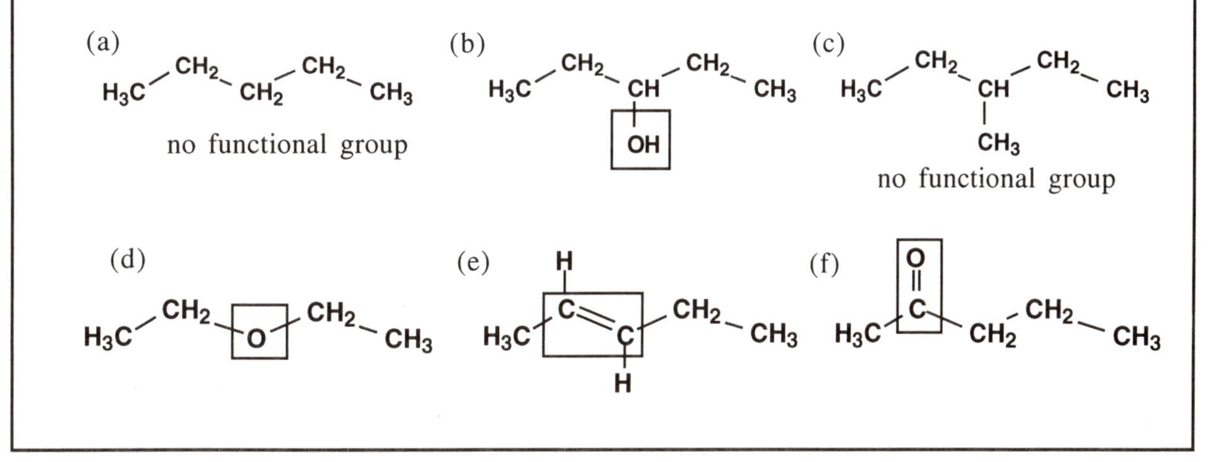

Repeat the above exercise with the following structures.

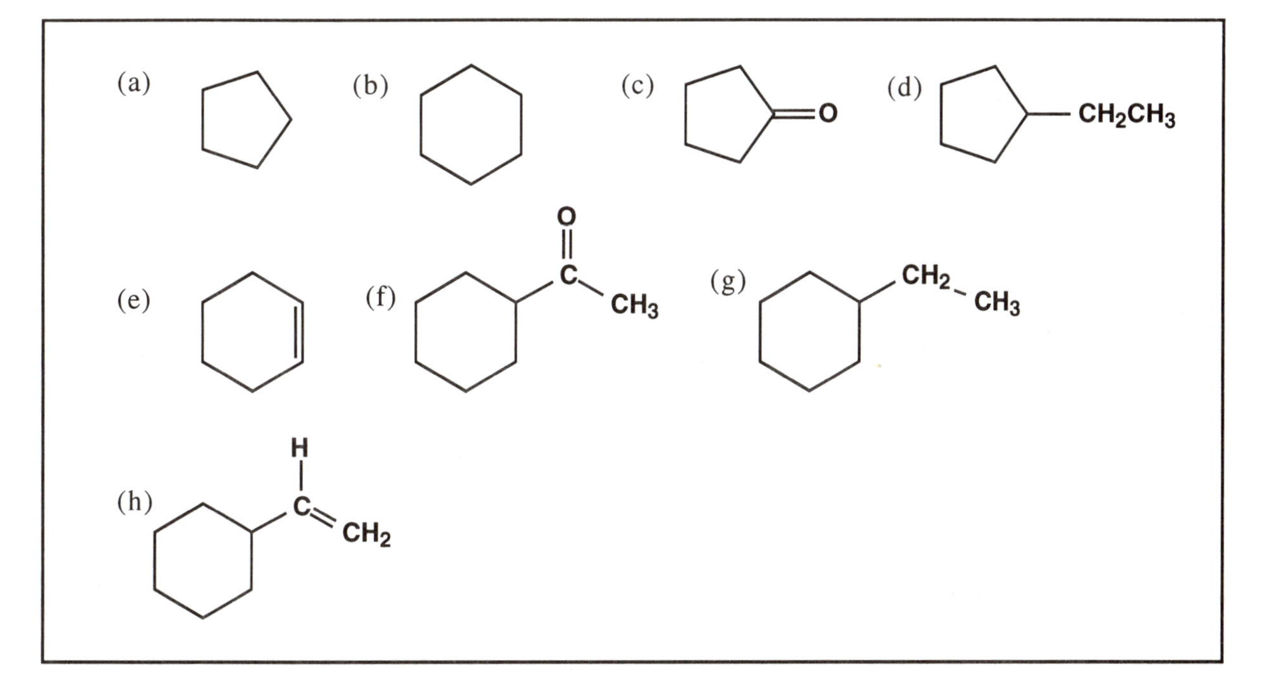

Answer

All the structures above are cyclic compounds but the principle is the same. Only the portions of the molecule which are not made up of C–C or C–H single bonds are functional groups. Therefore, structures (c), (e), (f) and (h) contain functional groups, as shown below.

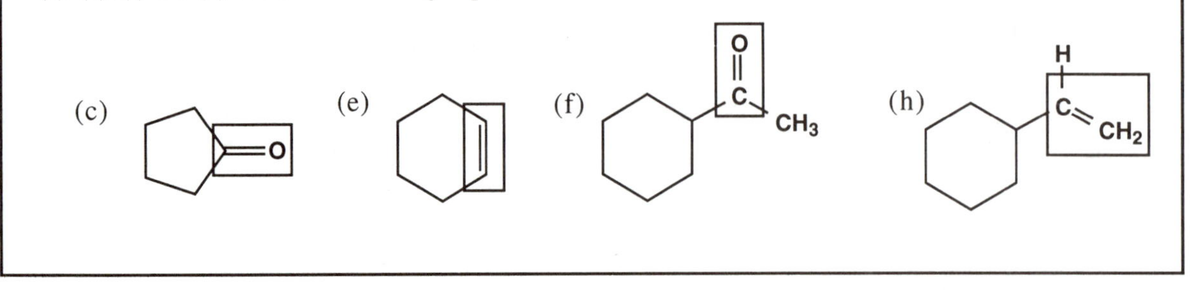

Molecules are not restricted to having just one functional group. It is possible to get a large number of functional groups on the one molecule. See if you can outline the functional groups in the following molecules.

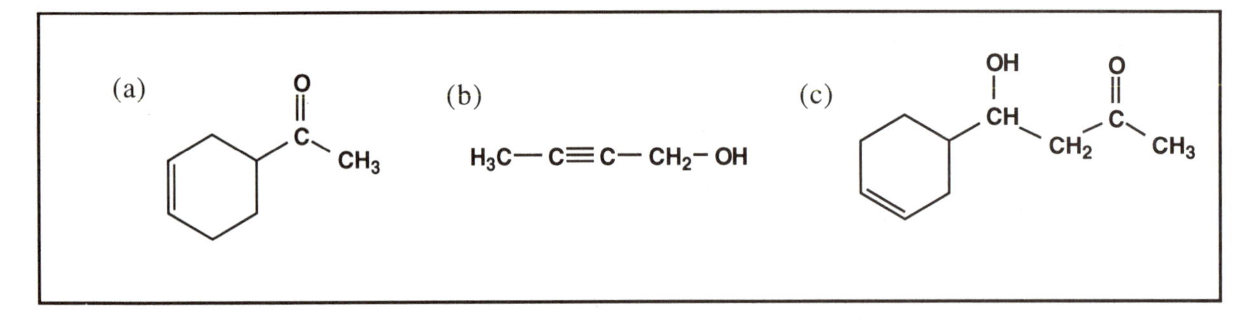

Answer

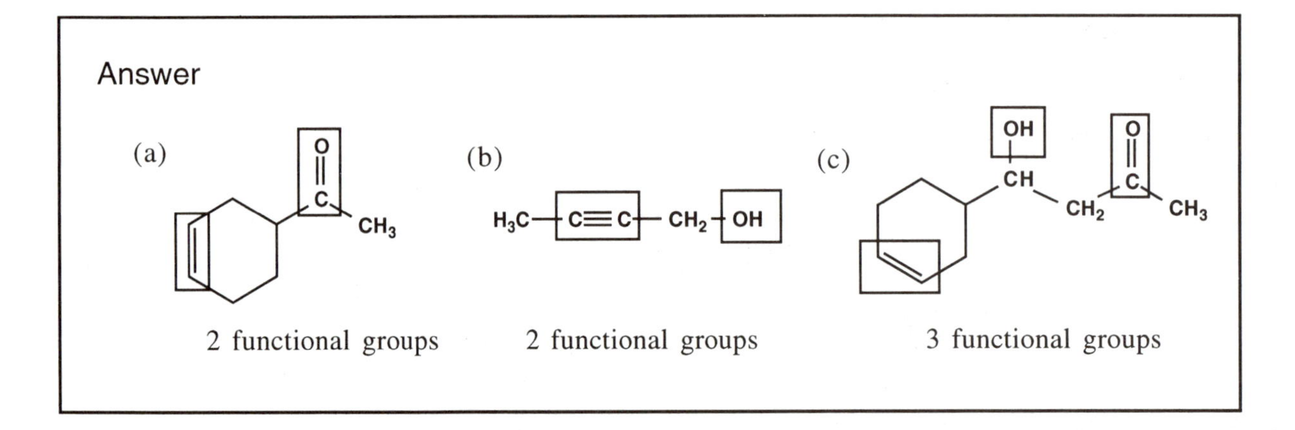

2.2 Naming functional groups

By now you should be able to pick out functional groups but you also need to name the functional group. This is important since the properties and reactions of any compound are determined by what type of functional group or groups it contains. The following are the most common functional groups.

- Functional groups which contain carbon and hydrogen only

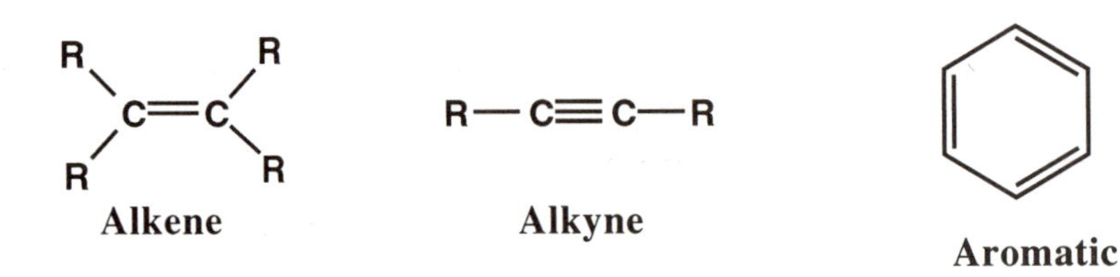

Alkene **Alkyne** **Aromatic**

- Functional groups which contain nitrogen

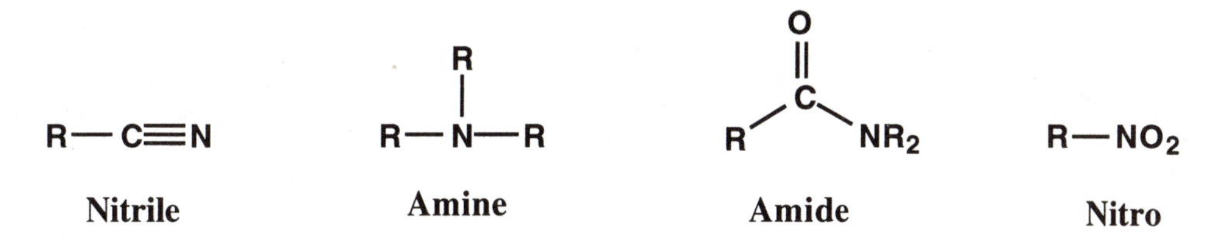

Nitrile **Amine** **Amide** **Nitro**

- Functional groups involving single bonds and which contain oxygen

R—OH

Alcohol or alkanol

Ar—OH

Phenol
(Ar = aromatic ring)

Ether

- Functional groups involving double bonds and which contain oxygen

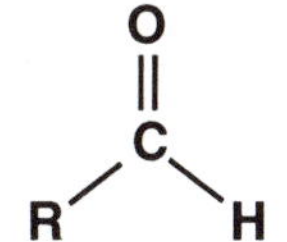

Aldehyde or alkanal

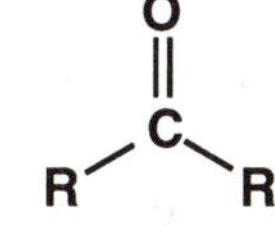

Ketone or alkanone

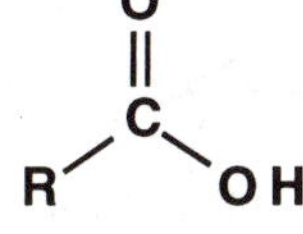

Carboxylic acid

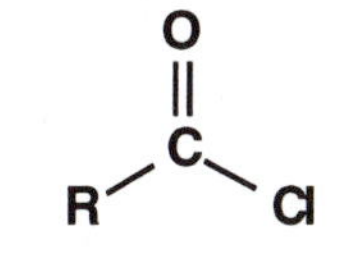

Acid chloride

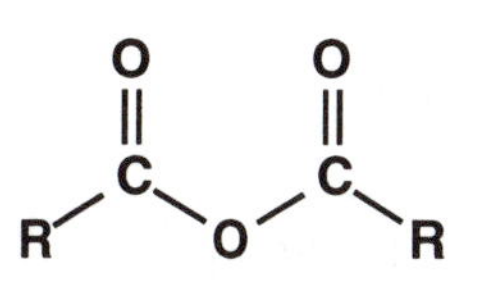

Acid anhydride

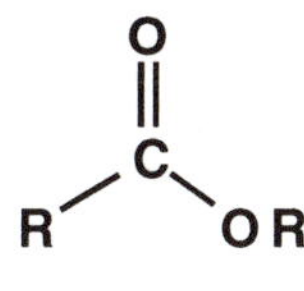

Ester

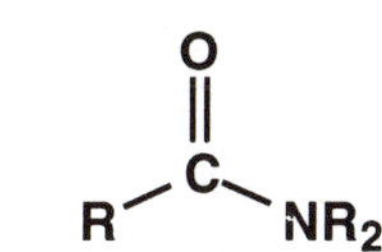

Amide

- Functional groups which contain a halogen atom

R—X (where X = F, Cl, Br or I)

Alkyl halide or haloalkane

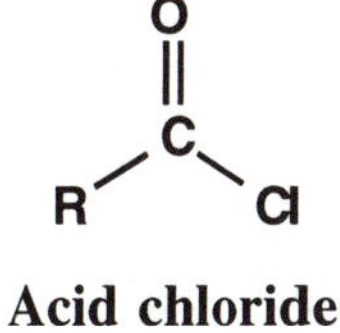

Acid chloride

2.3 Functional groups containing carbon and hydrogen only

The following molecules have functional groups containing carbon and hydrogen atoms only. Outline and identify the functional groups.

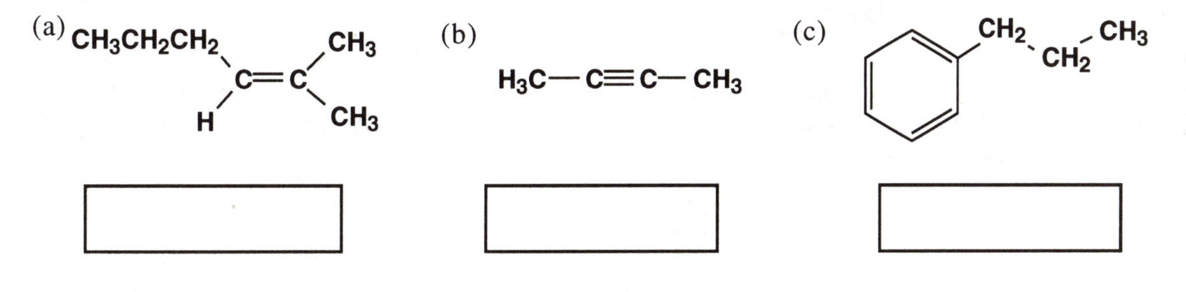

Answer

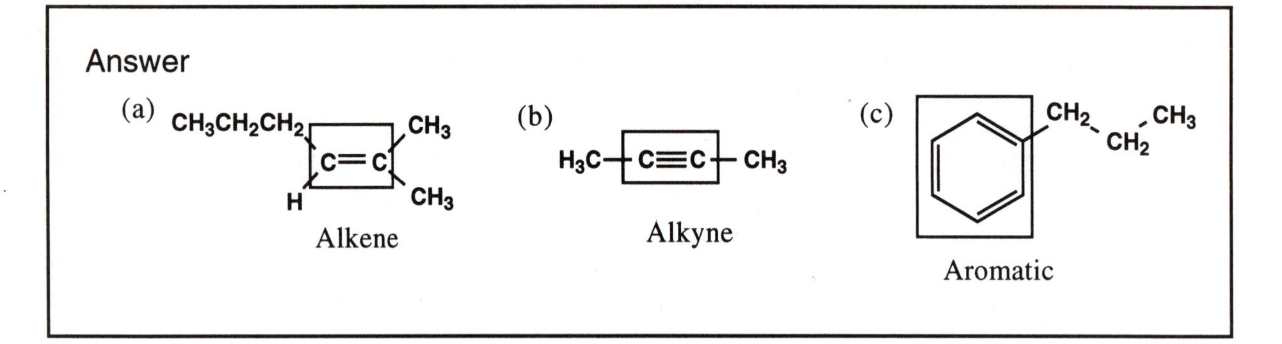

The compound below was given in an exam and students were asked to identify the functional group(s). Several students gave the answer shown but were marked wrong. Why?

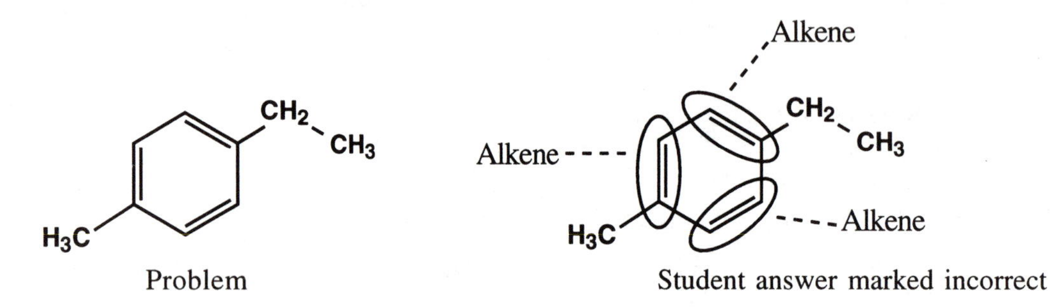

Answer

The answer given was wrong since the functional group should have been the aromatic ring as shown below.

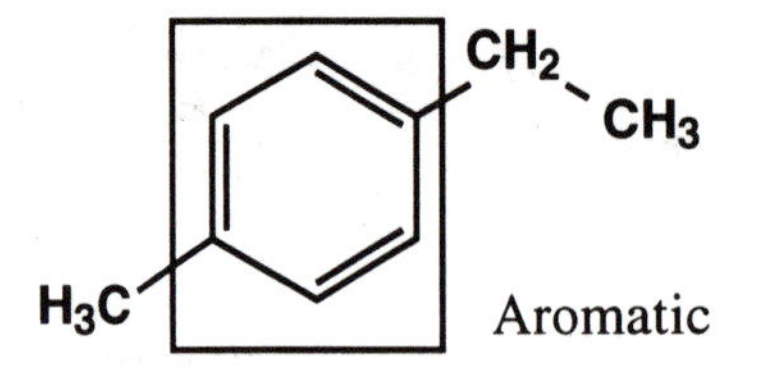

The diagram used for an aromatic ring uses three double bonds but this is just a representation for the aromatic ring. The π electrons in an aromatic ring are delocalised round the ring. The aromatic ring can also be represented as below.

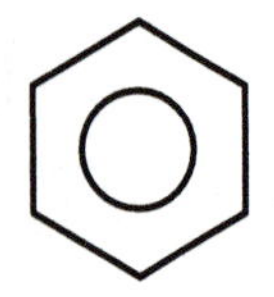

The circle in the centre represents the delocalisation of six π electrons.

Outline and identify the functional groups on the following molecule

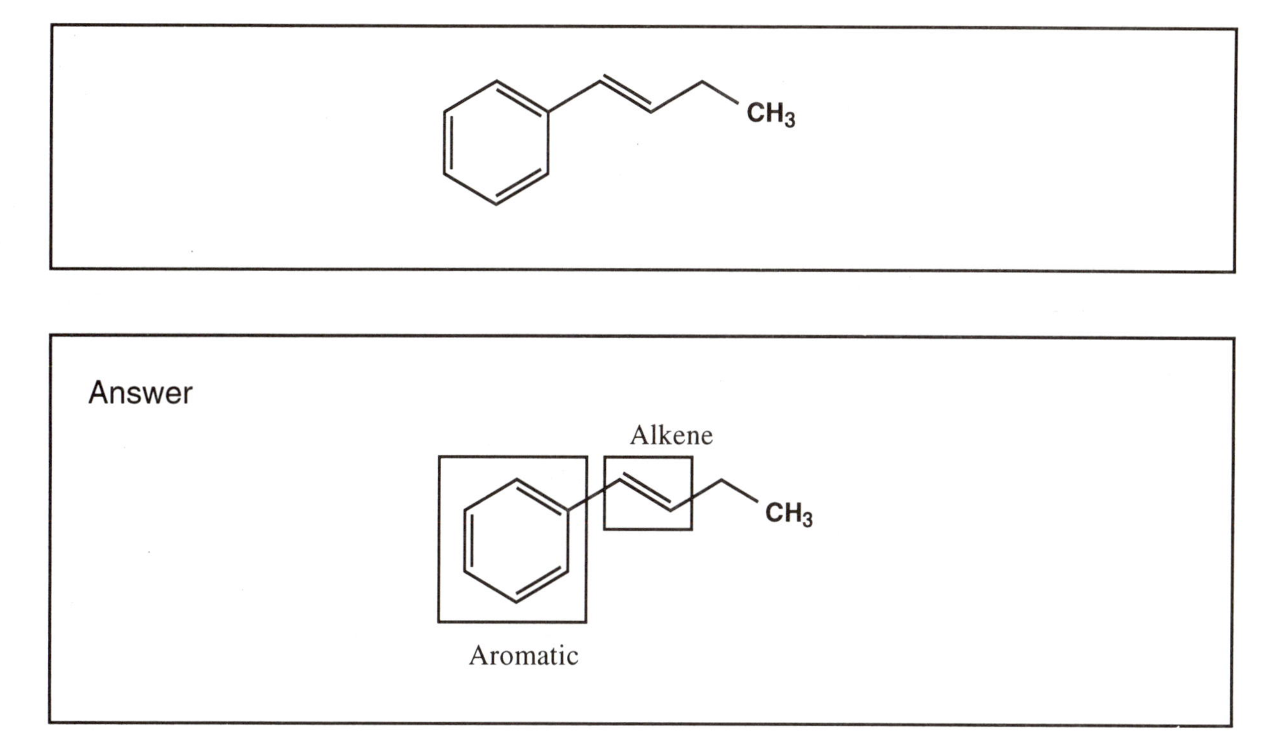

2.4 Functional groups which contain nitrogen

The following molecules have a functional group containing nitrogen. Outline and identify the functional groups.

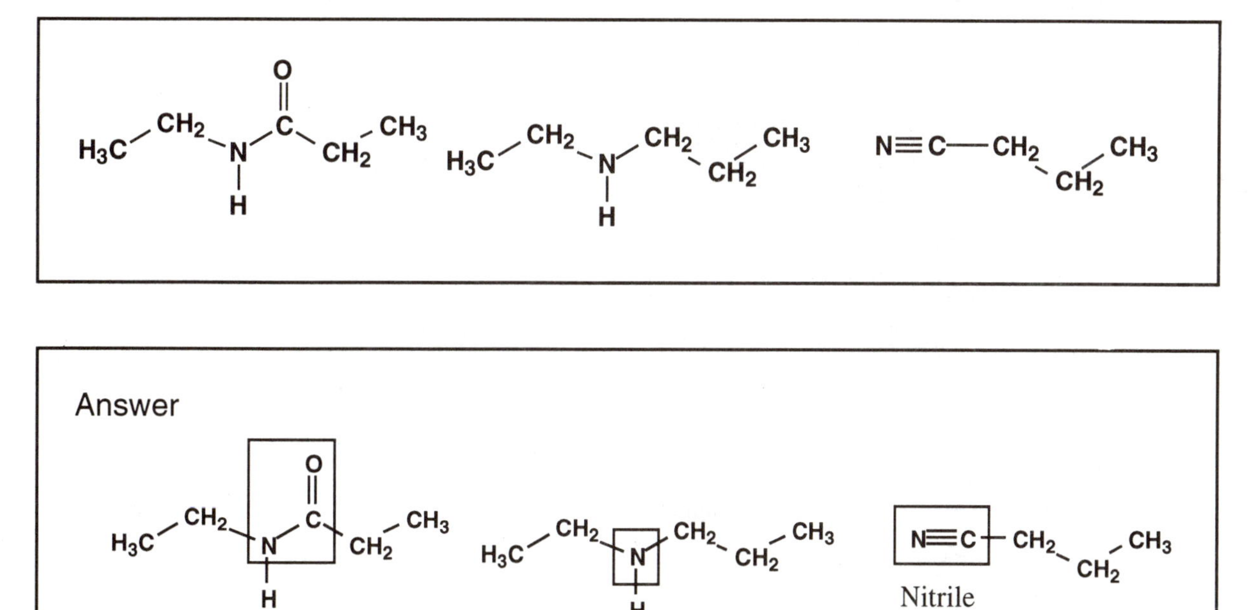

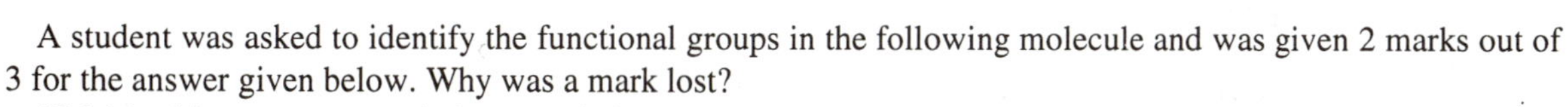

A student was asked to identify the functional groups in the following molecule and was given 2 marks out of 3 for the answer given below. Why was a mark lost?

('Me' in this structure stands for a methyl group ($-CH_3$). This expression is often used to represent individual methyl groups.)

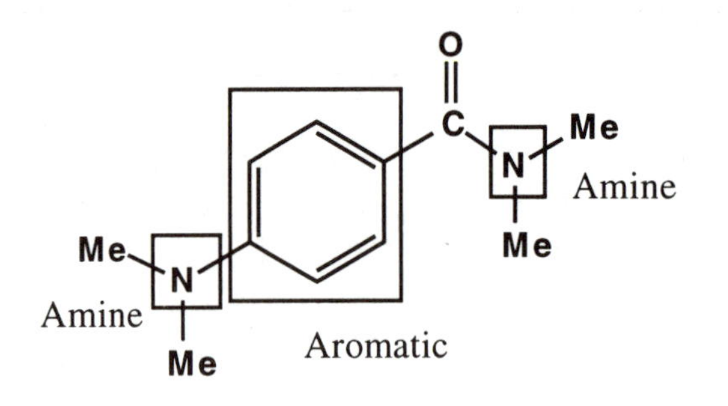

Answer

The student failed to identify the amide. It is a common mistake to see a nitrogen atom and to assume that this always represents an amine. You must look to see whether a carbonyl group (C=O) is attached to the nitrogen. If there is, then the functional group is an amide and not an amine.

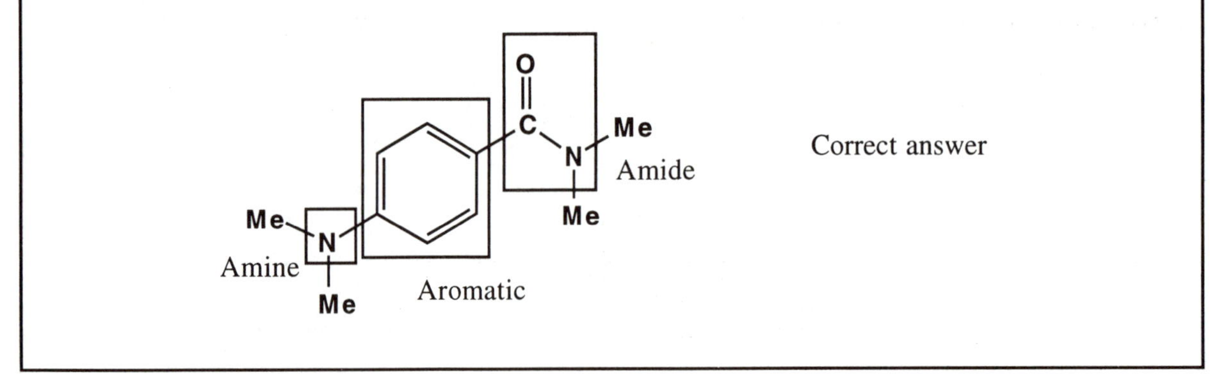

2.5 Functional groups which contain oxygen singly bonded to carbon

We now move to functional groups which contain an oxygen. Outline and identify the functional groups in the following molecules.

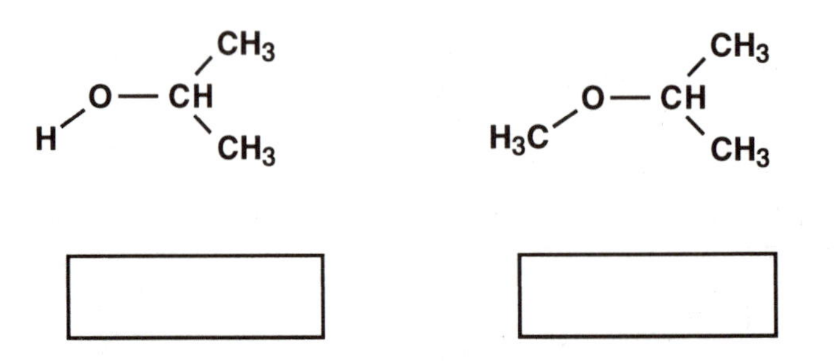

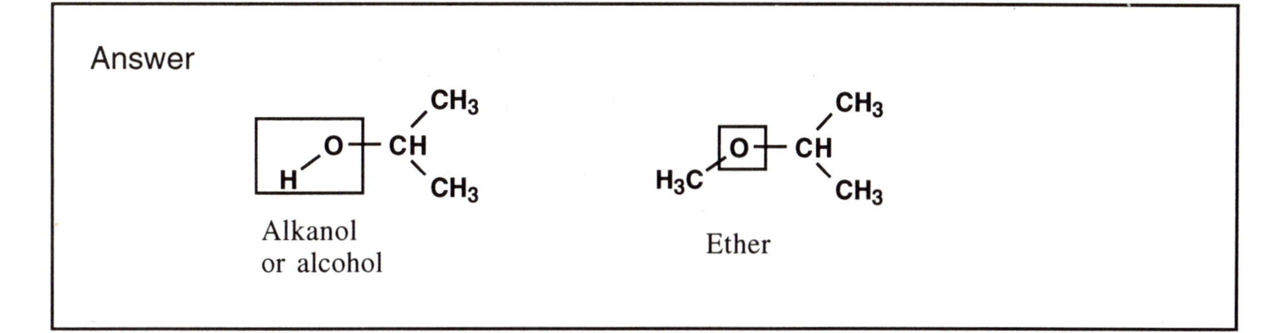

Outline and identify the functional group in the following molecule.

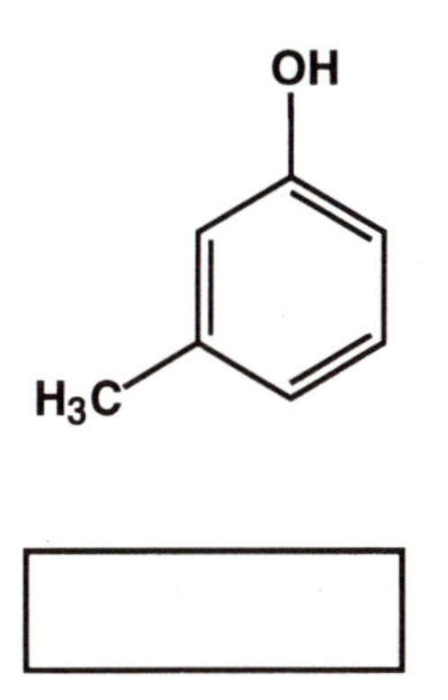

Answer

This structure is a phenol. The functional group is the aromatic ring *and* the OH group.

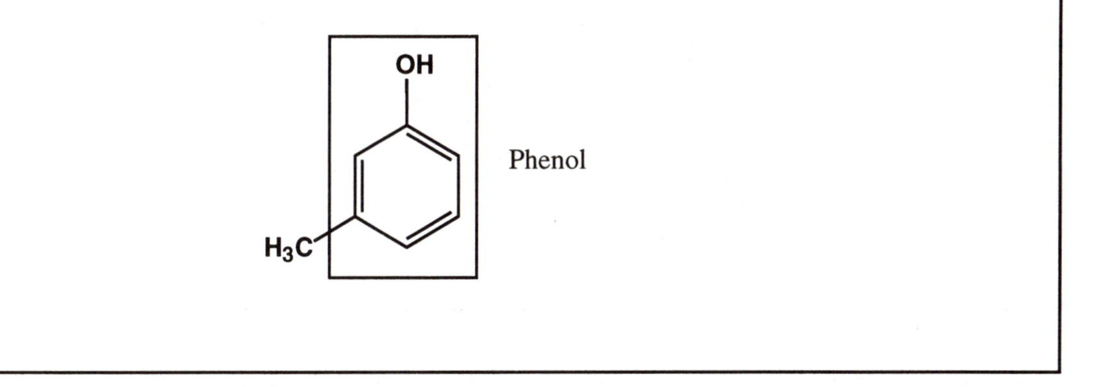

A student was asked in an examination to identify the functional groups in the following molecule. The answer given is shown below and was given half marks. Why?

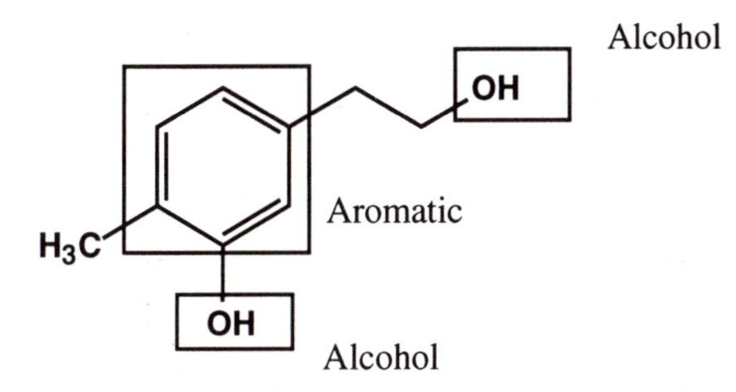

Answer

The correct answer is given below. The student identified both OH groups as alcohols but only one of them is an alcohol. The OH group directly attached to the aromatic ring is part of a phenol and has different chemical properties from an alcohol functional group.

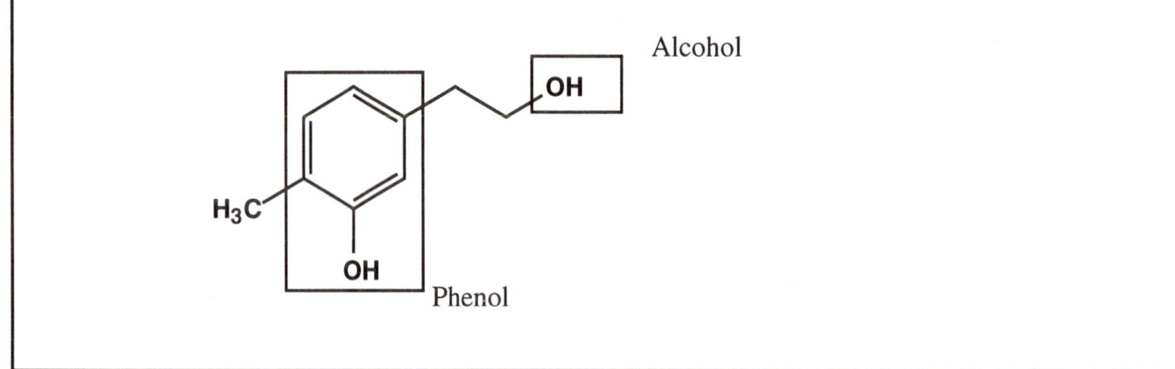

2.6 Functional groups which contain oxygen doubly bonded to carbon

The functional groups above involve an oxygen bonded to two separate groups by single bonds.
The following compounds contain an oxygen bonded to carbon by a double bond. Outline and identify the functional groups.

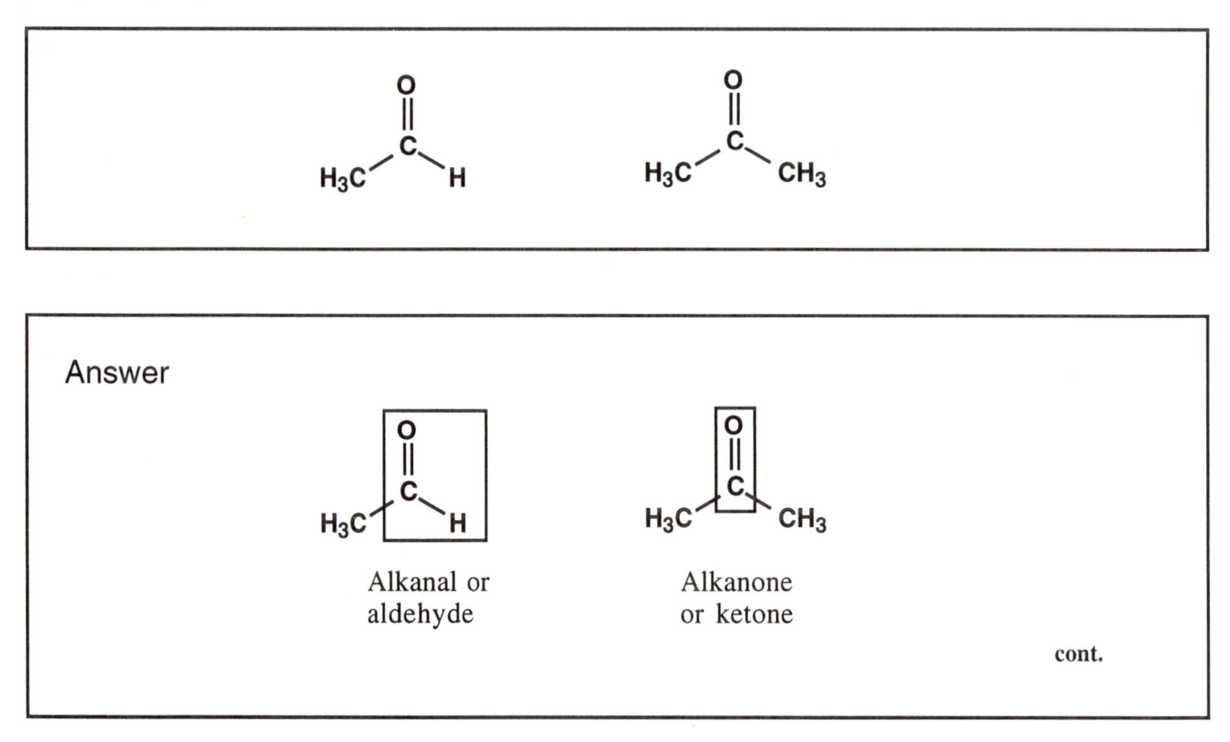

When it comes to recognising functional groups involving C=O, it is vital to look at what is next to the C=O group. If alkyl (or aryl) groups are on both sides it is a ketone. If there is an alkyl (or aryl) group on one side and a proton on the other side then it is an aldehyde.

The C=O group is known as a **carbonyl group** in both aldehydes and ketones.

The following structures are aldehydes or ketones. Identify which are which and circle the aldehyde or ketone functional group.

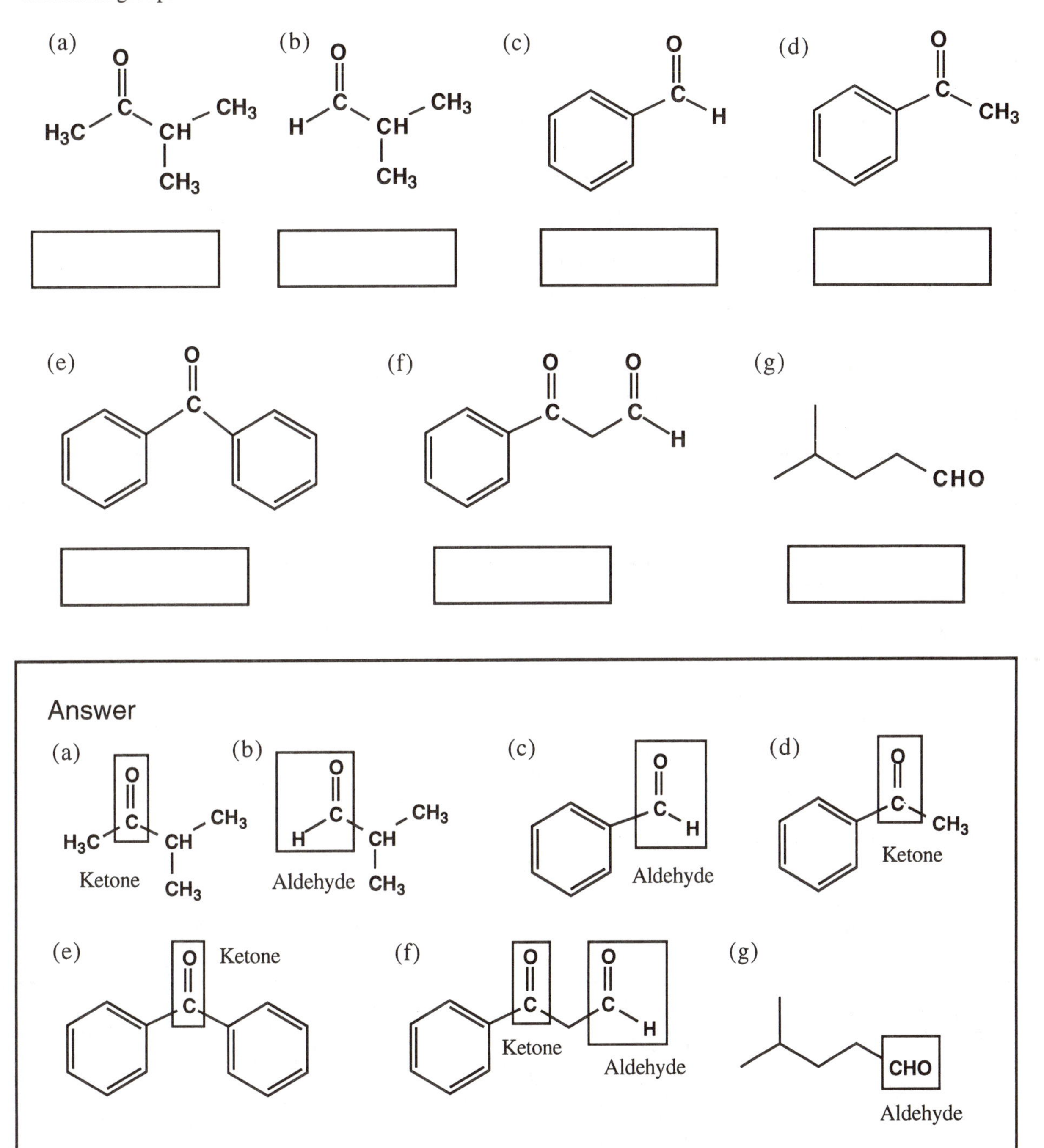

Notice the aldehyde functional group in example (g). Structure (g) is a simplified method of representing the aldehyde functional group, but it can sometimes be difficult to recognise as such. Remember that carbon must always have four bonds. Therefore, the CHO group must have a C=O or carbonyl group as well as single bonds to hydrogen and the rest of the molecule.

If the ketone or aldehyde has alkyl groups *directly* attached then it is known as an aliphatic ketone or aldehyde. If one or more aromatic rings are *directly* attached then it is an aromatic aldehyde or ketone. Identify the aromatic and aliphatic aldehydes and ketones in the above examples.

Answer

Structure (a) is an aliphatic ketone which obeys the general formula R–(CO)–R.
Structures (b) and (g) are aliphatic aldehydes which obey the general formula R–CHO.
Structure (c) is an aromatic aldehyde which obeys the general formula Ar–CHO.
Structures (d) and (e) are aromatic ketones which obey the general formula Ar–(CO)–(R or Ar).
Structure (f) is an aromatic ketone but also contains an aliphatic aldehyde since the aromatic group is not directly attached to the aldehyde group.

Outline and identify the functional group in the following structure and explain whether the group is an aliphatic or aromatic group.

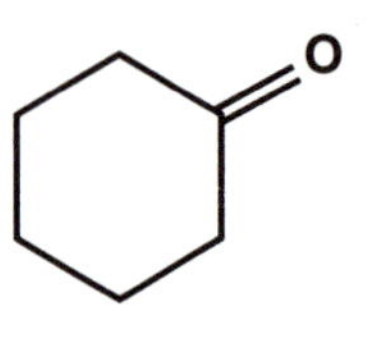

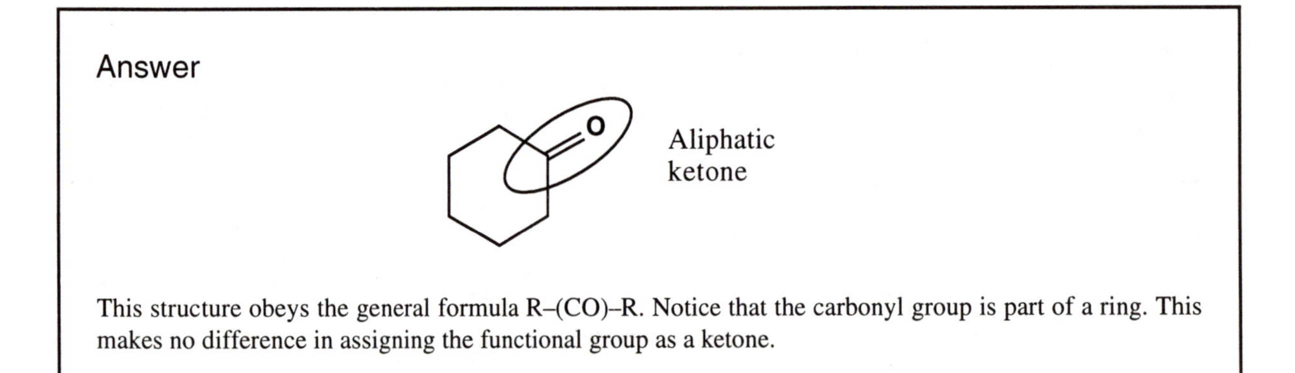

This structure obeys the general formula R–(CO)–R. Notice that the carbonyl group is part of a ring. This makes no difference in assigning the functional group as a ketone.

One common mistake made when starting organic chemistry is to assume that any six-membered ring is aromatic. Remember that an aromatic ring is represented by either of the structures

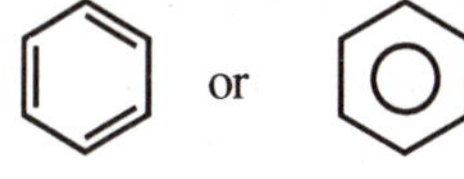

If you called the above structure an aromatic ketone you were wrong. In fact, it would be impossible to have a carbonyl group as part of an aromatic ring. Why?

Answer

The structure below is impossible since there is a carbon with five bonds.

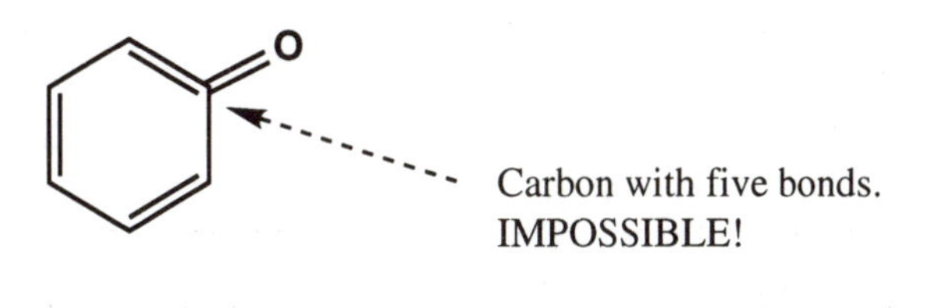

We have covered aldehydes and ketones, but there are other types of functional groups which contain the C=O group. These are the carboxylic acids and their derivatives. The chemical properties of carboxylic acids and their derivatives are different from aldehydes and ketones. Outline and identify the functional groups in the following molecules.

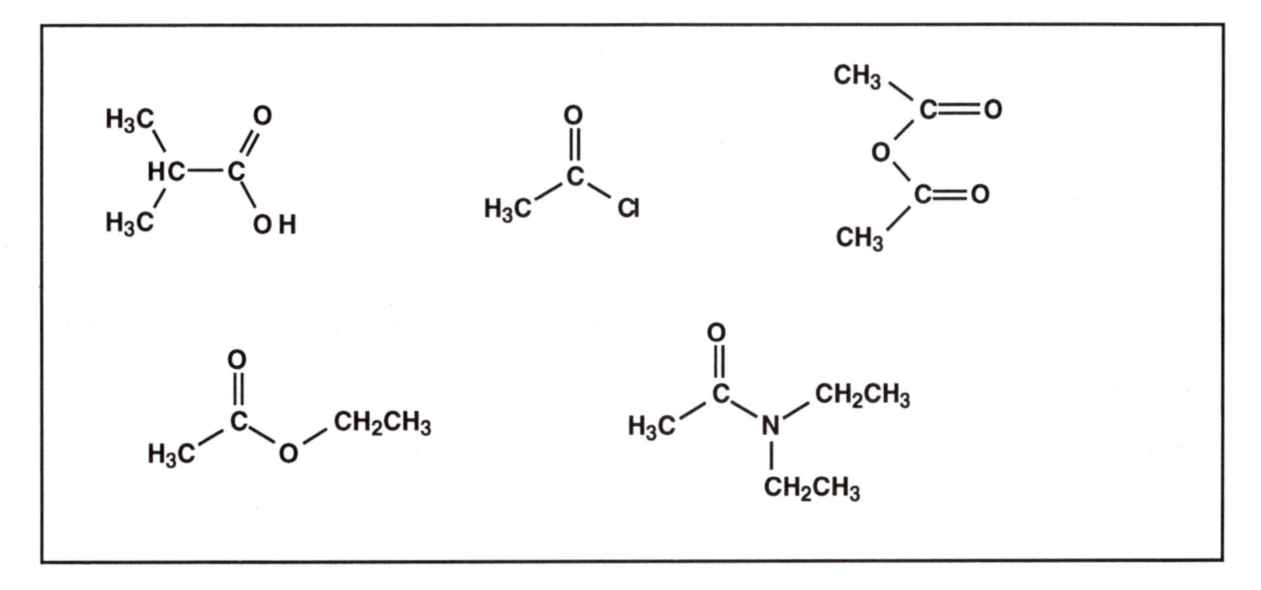

Answer

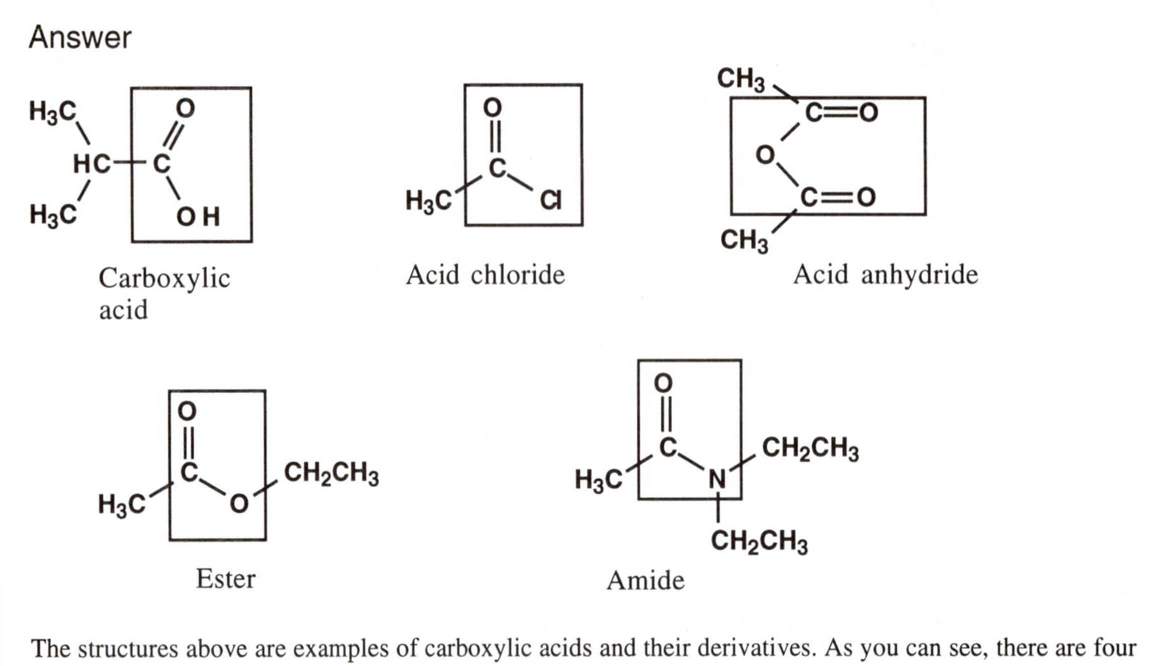

The structures above are examples of carboxylic acids and their derivatives. As you can see, there are four derivatives. They are called derivatives because it is possible to synthesise these derivatives from a carboxylic acid. Also, all carboxylic acid derivatives can be hydrolysed back to their parent carboxylic acids.

Sometimes you will see structures like the ones below which simplify the functional group. Outline and identify the functional groups and state whether they are aromatic or aliphatic. (Remember that the aromatic ring must be directly next to the functional group if it is to be aromatic.)

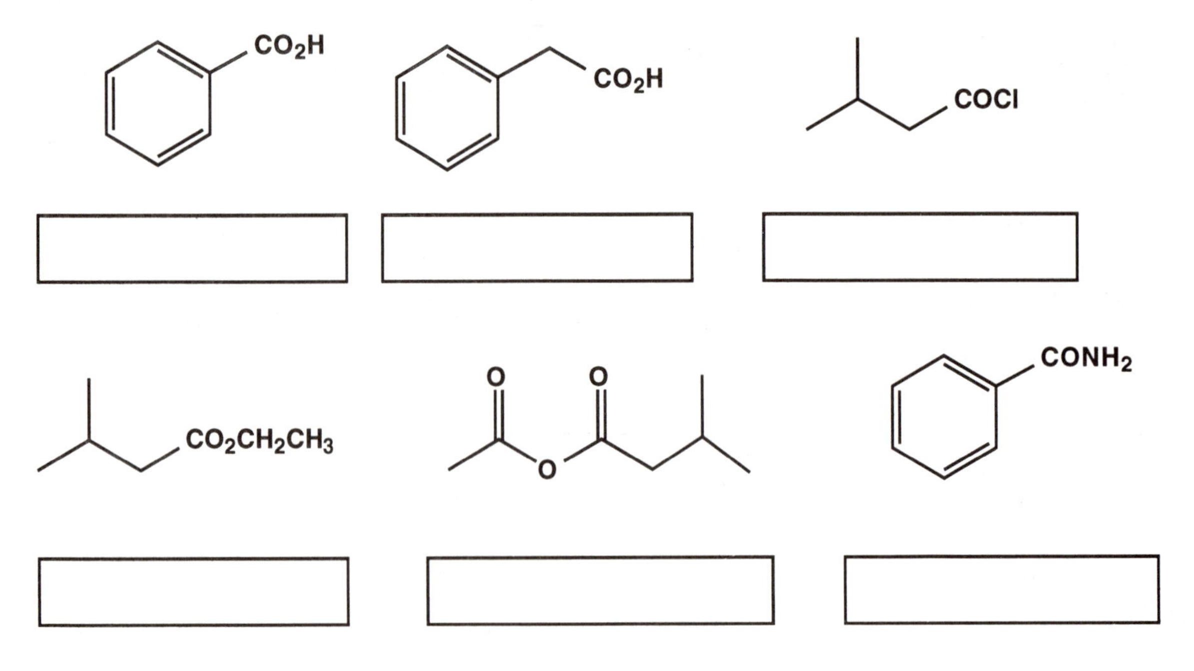

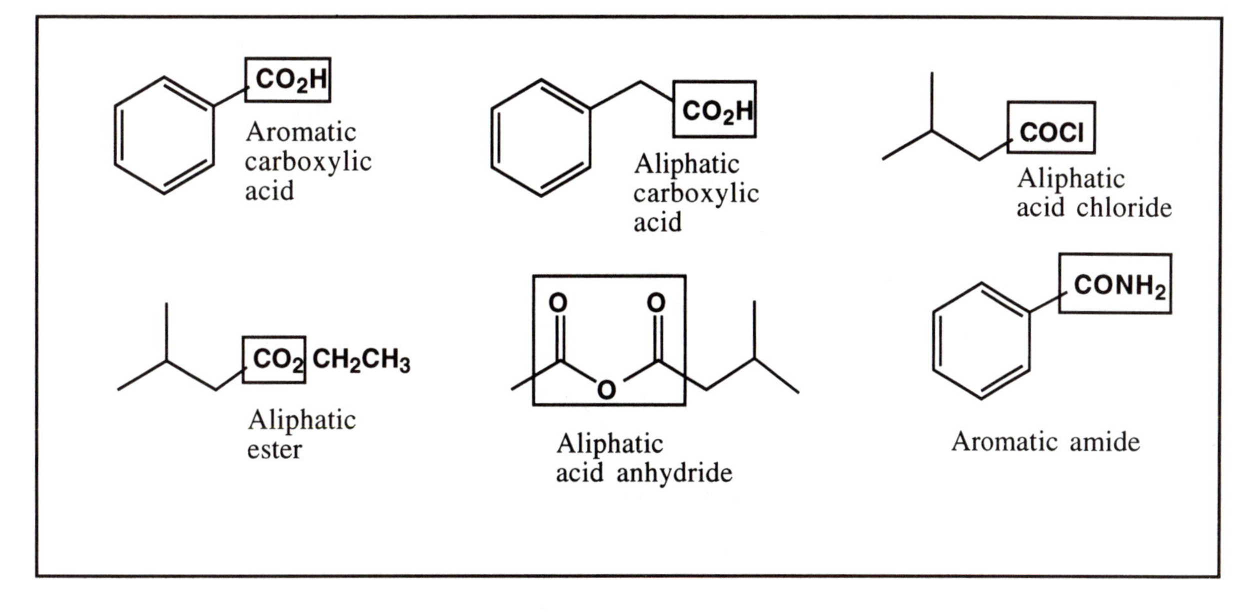

At this stage it is worth pointing out that an acid derivative is defined as being aromatic or aliphatic depending on whether there is an aryl group **directly** attached to the **carbonyl** end of the functional group, that is

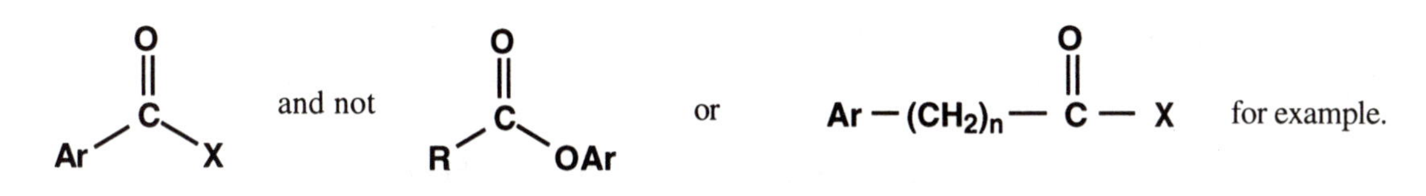

Try the following structures. Outline and identify the functional groups and state whether they are aromatic or aliphatic. Note: Ph (shorthand for phenyl) is frequently used as shorthand for an aromatic ring. For example, $MeCO_2Ph$ represents the following structure:

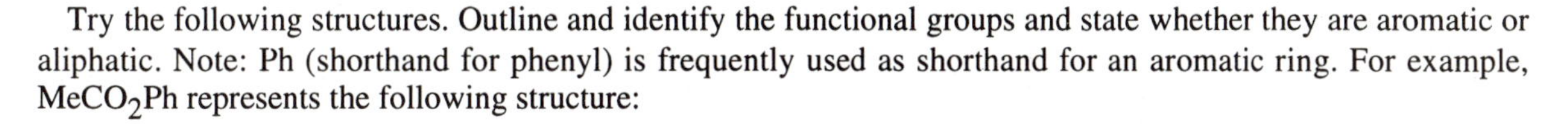

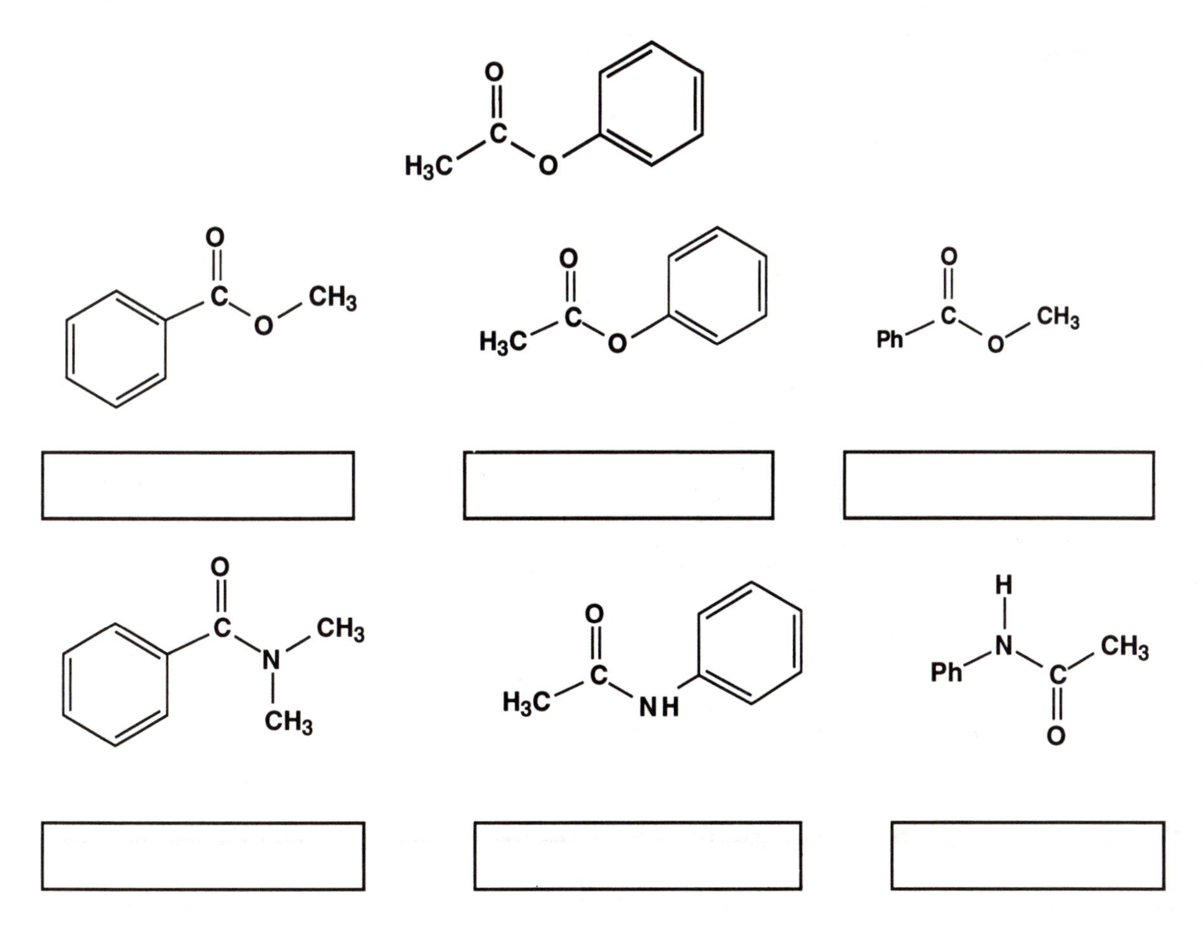

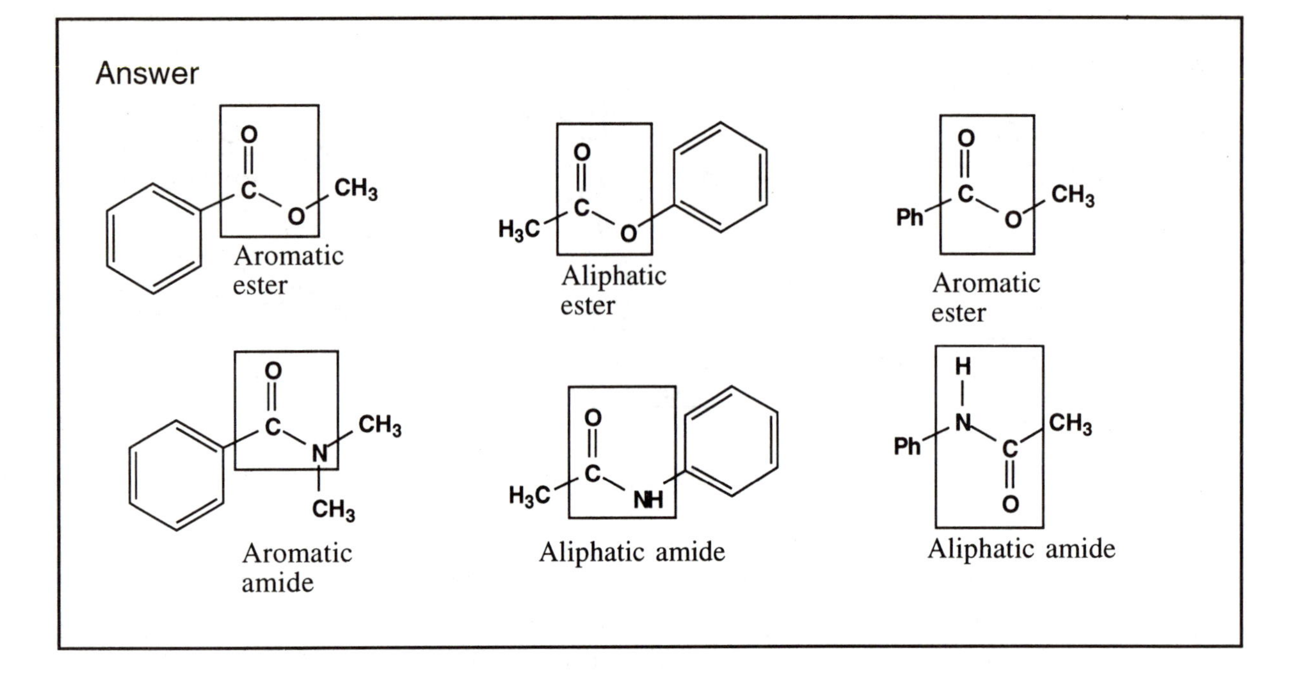

2.7 Functional groups which contain a halide atom

There are three types of functional groups which contain a halide atom (X)—acid halides, alkyl halides and aryl halides.

The general formulae for these are as follows. Notice that the acyl chloride can be aromatic or aliphatic depending on the group attached to the carbonyl group.

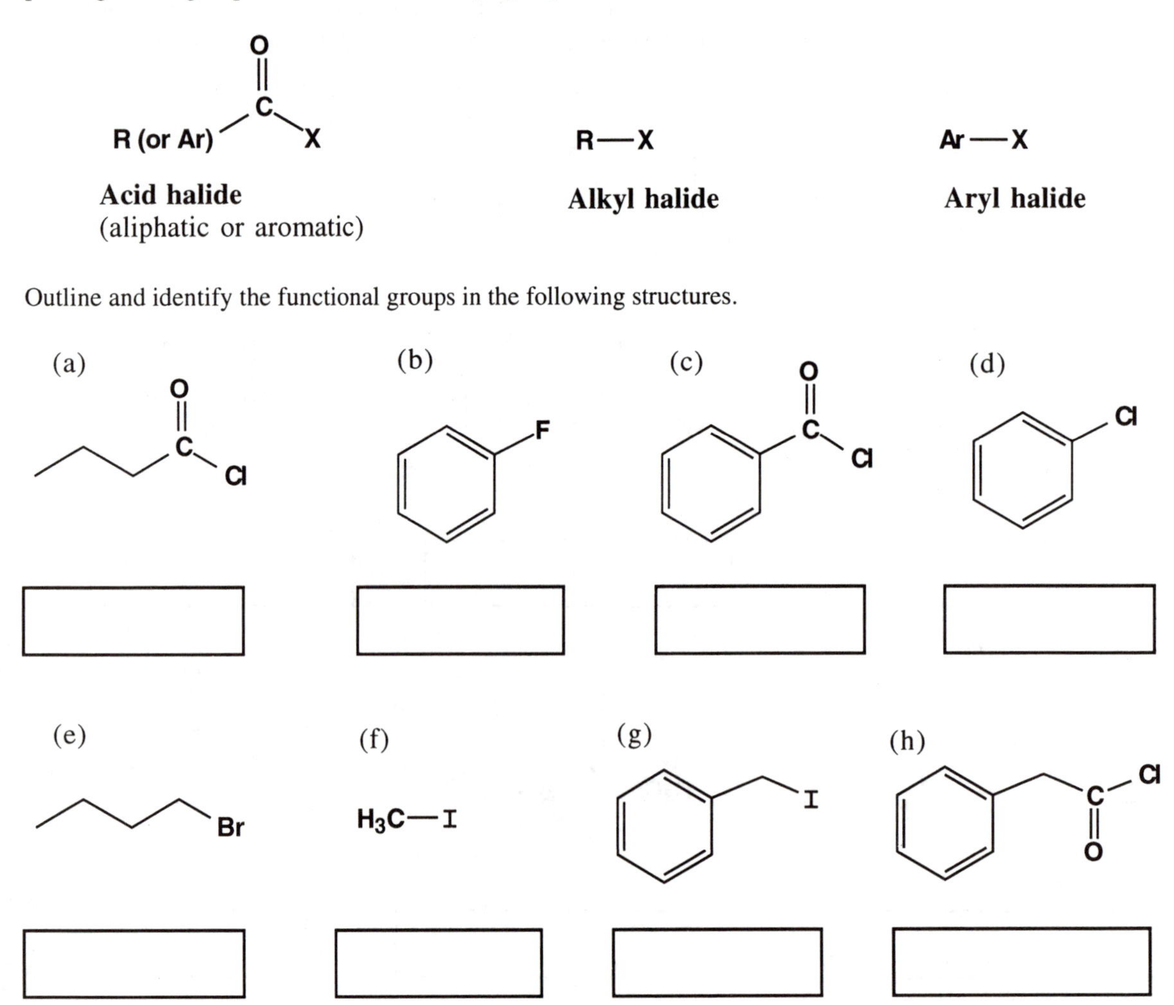

Outline and identify the functional groups in the following structures.

Answer

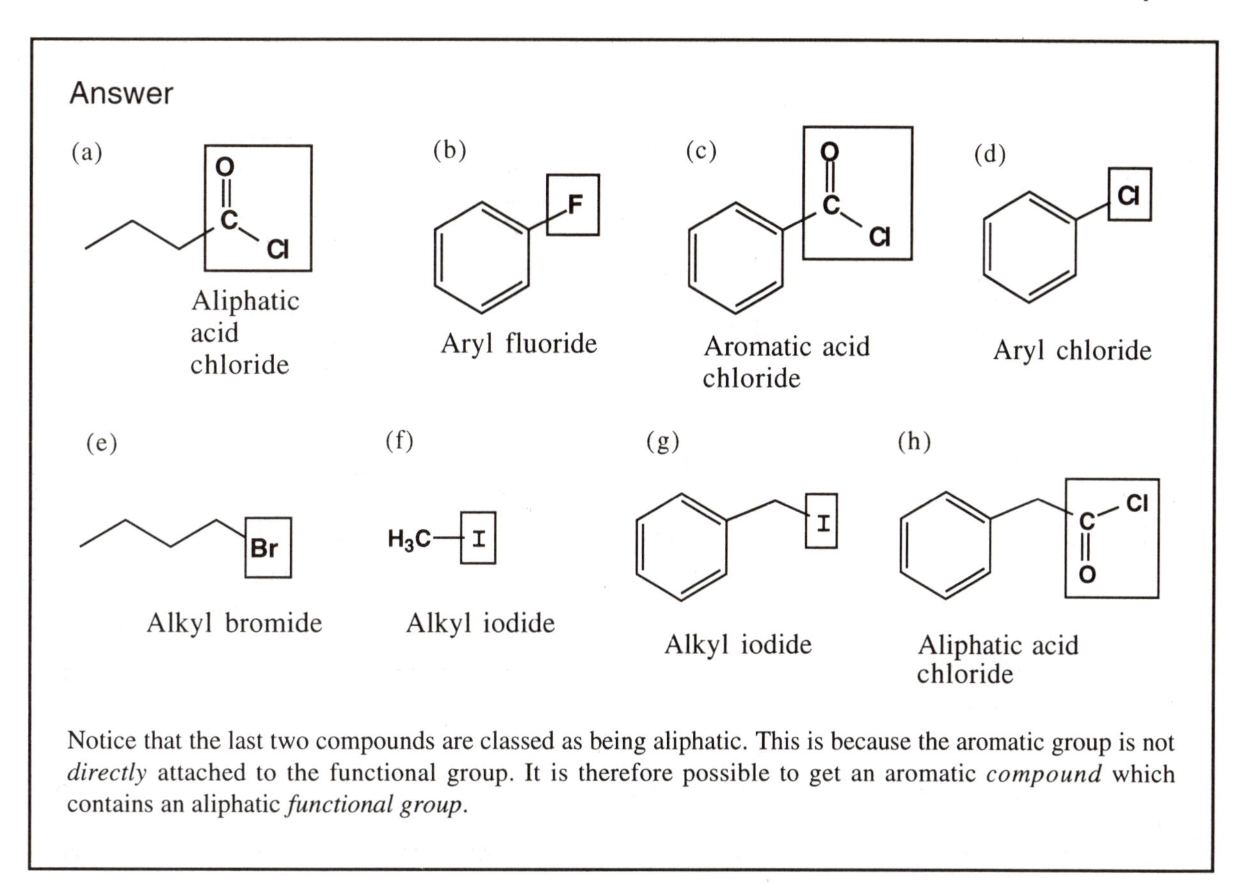

Notice that the last two compounds are classed as being aliphatic. This is because the aromatic group is not *directly* attached to the functional group. It is therefore possible to get an aromatic *compound* which contains an aliphatic *functional group.*

2.8 Compounds with more than one functional group

Many naturally occurring compounds contain more than one functional group. Half the battle in understanding the properties of these compounds is being able to identify the various functional groups. Outline and identify the functional groups in the following important biologically active compounds.

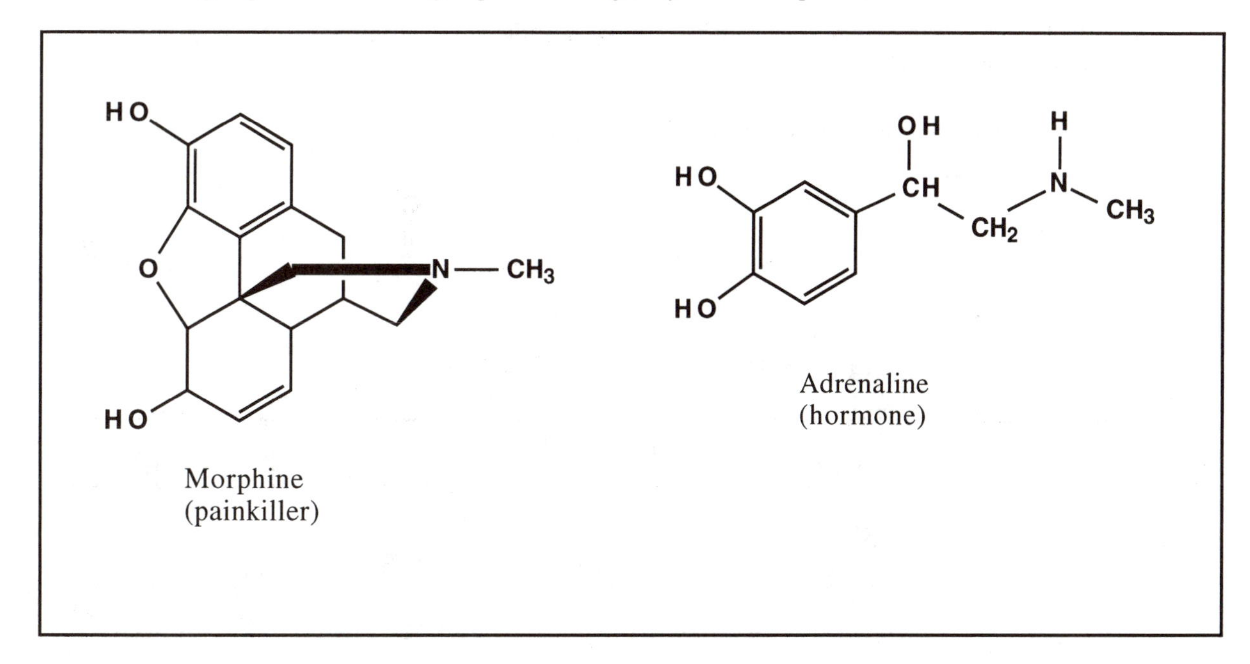

Answer

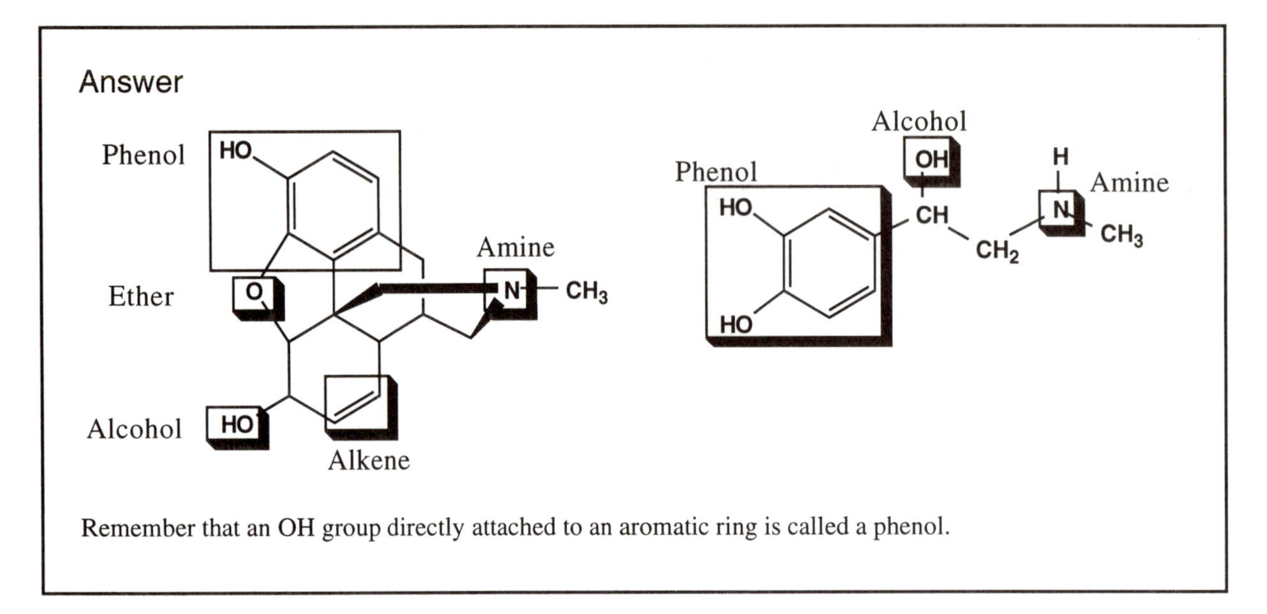

Remember that an OH group directly attached to an aromatic ring is called a phenol.

Identify the functional groups in the following natural products.

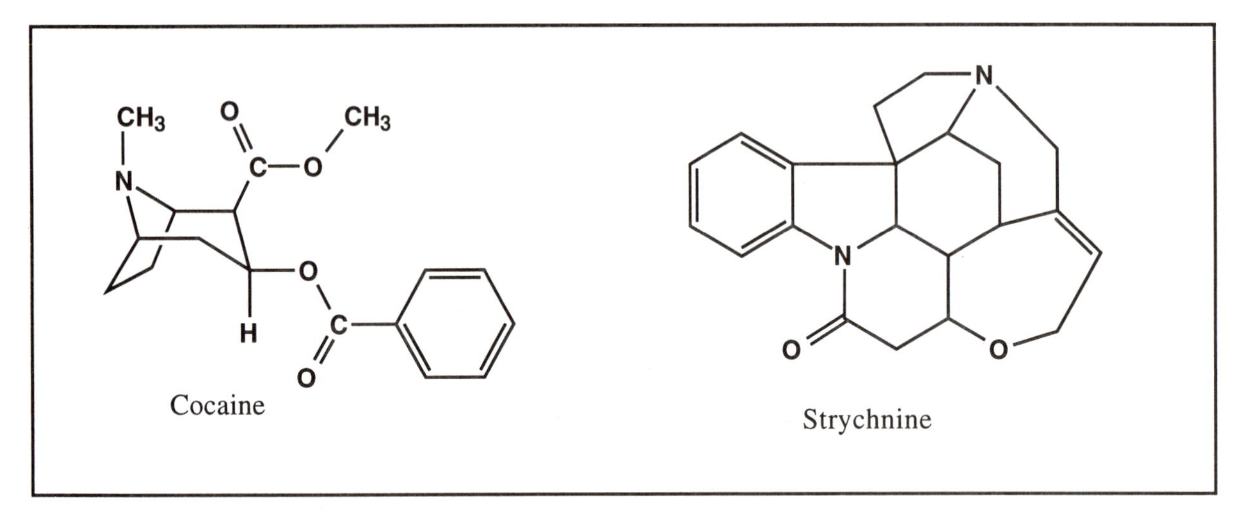

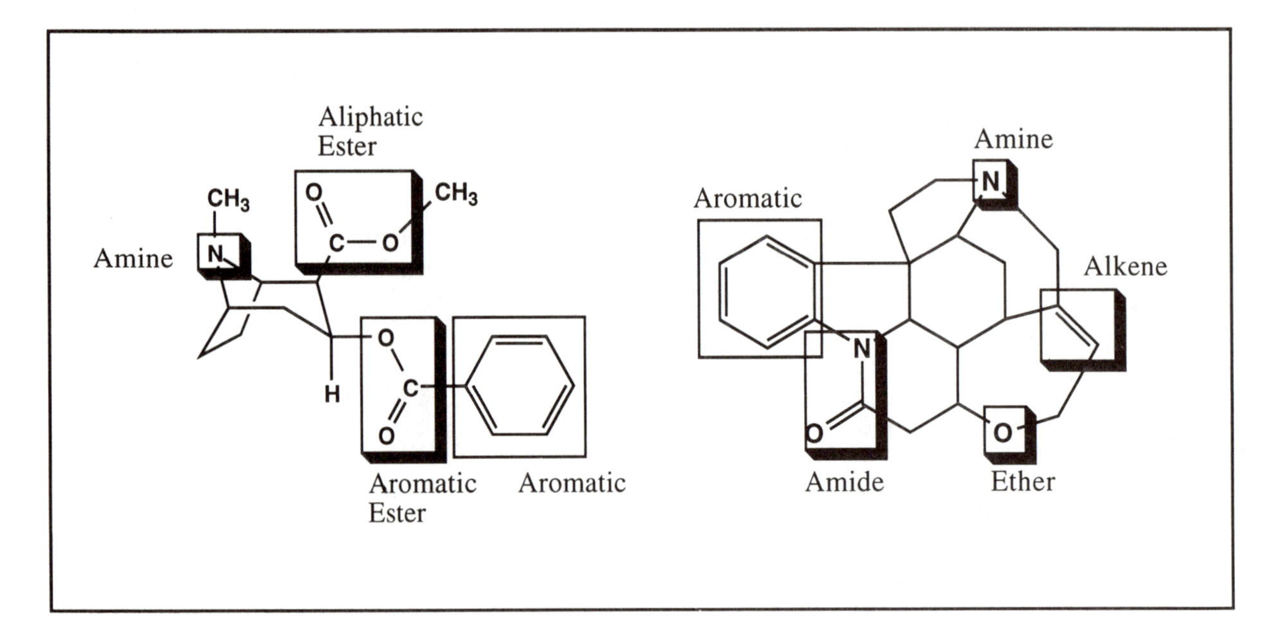

Summary

- Any portion of an organic molecule which is not made up of C–C single bonds or C–H single bonds is a functional group.
- The chemical properties of an organic compound are determined by the functional groups present.
- The most common functional groups are those which:
  - (a) contain carbon and hydrogen only—alkenes, alkynes, aromatic;
  - (b) contain nitrogen—amines, amides, nitriles;
  - (c) contain oxygen singly bonded to carbon—alkanols (alcohols), ethers, phenols
  - (d) contain oxygen doubly bonded to carbon—alkanals (aldehydes), alkanones (ketones), carboxylic acids, acid chlorides, acid anhydrides, esters, amides;
  - (e) contain a halide atom—alkyl halides, acyl halides, acid chlorides.

SECTION 3

Formulae and Molecular Weights

Formulae and Molecular Weights

Now that we have looked at functional groups, we shall look at some simplified structures containing functional groups to check that you can work out molecular formulae and weights.

Give the molecular formulae and molecular weights of the following.

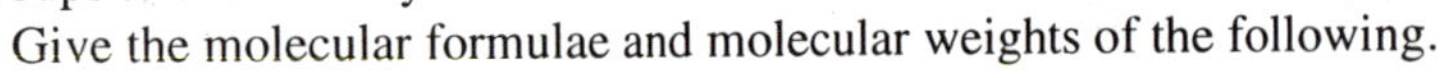

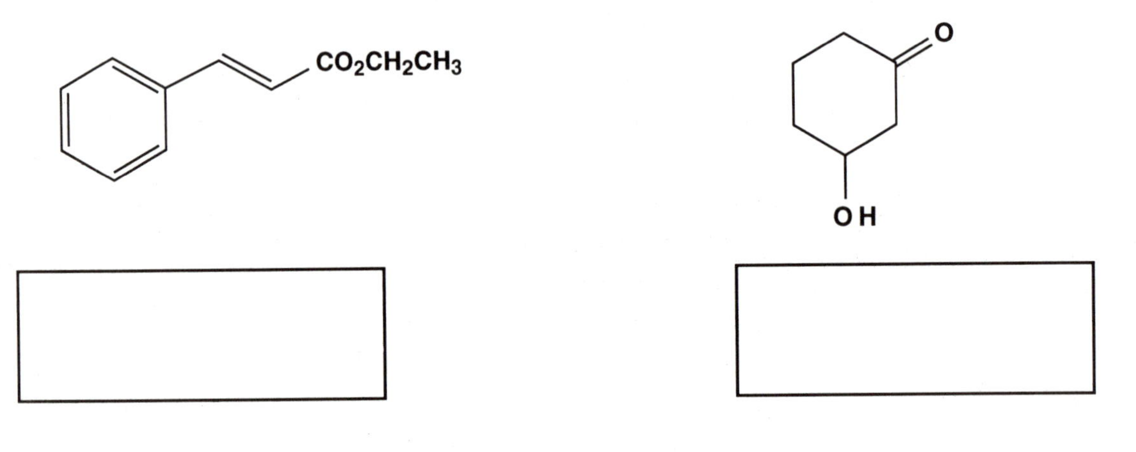

Answer

We shall put in all the atoms in case you missed any hydrogens.

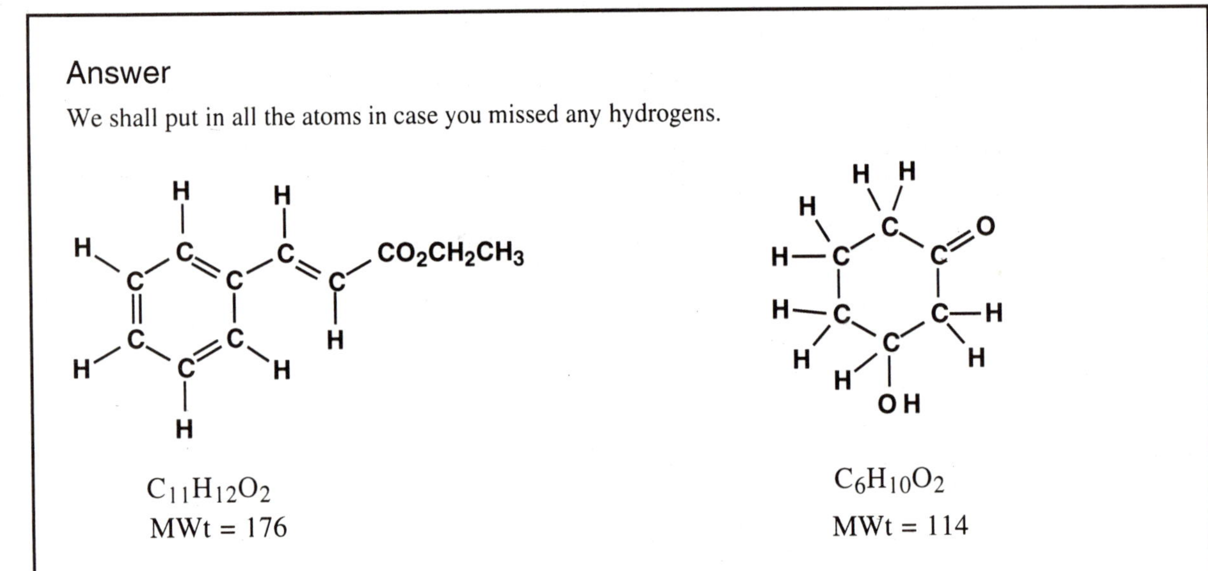

$C_{11}H_{12}O_2$
MWt = 176

$C_6H_{10}O_2$
MWt = 114

Give the molecular formulae and weights for the following structures.

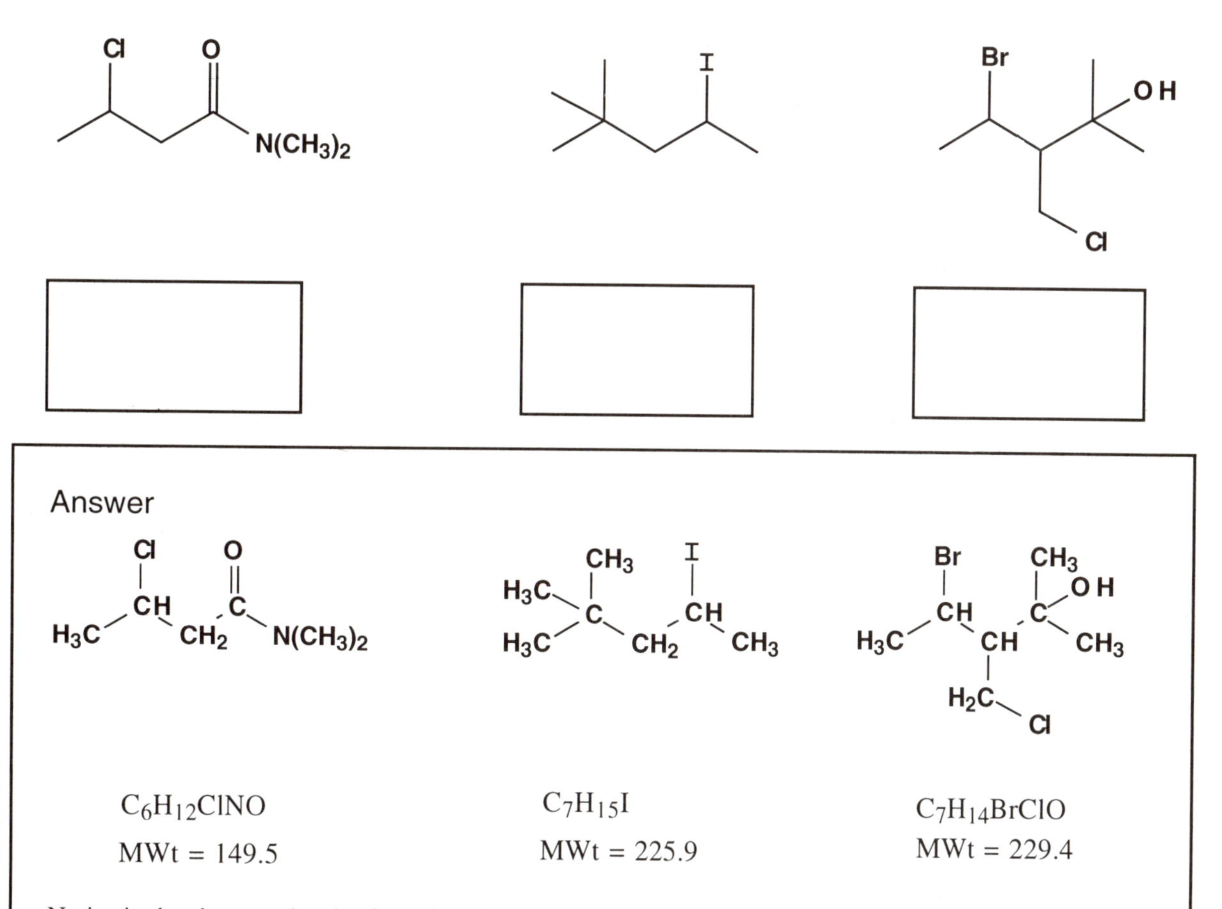

Notice in the above molecular formulae that the carbon and hydrogen come first, followed by the other elements in alphabetical order.

This is the conventional way of expressing molecular formulae. By having a consistent rule such as this, it is possible to have molecular formula indices in reference books and catalogues, making the search for a particular compound much easier.

SECTION 4

Nomenclature—Alkanes

Nomenclature—Alkanes

4.1 Staight-chain alkanes

It is important to identify chemicals by their correct names. Otherwise, chaos would reign. The brewer putting ethan**al** into his ale rather than ethan**ol** would not be very popular with his customers. Ethanal and ethanol are two totally different compounds with different properties, not least in taste! There are procedures to be followed when naming organic compounds and we shall consider these by looking first at alkanes. Alkanes are the easiest organic compounds to name and so we shall master the rules for these before moving onto structures with functional groups.

You should already know how to name the following alkanes. Try your luck!

(a) CH_4 (b) CH_3CH_3 (c) $CH_3CH_2CH_3$ (d) $CH_3CH_2CH_2CH_3$ (e) $CH_3CH_2CH_2CH_2CH_3$

(f) $CH_3CH_2CH_2CH_2CH_2CH_3$ (g) $CH_3CH_2CH_2CH_2CH_2CH_2CH_3$

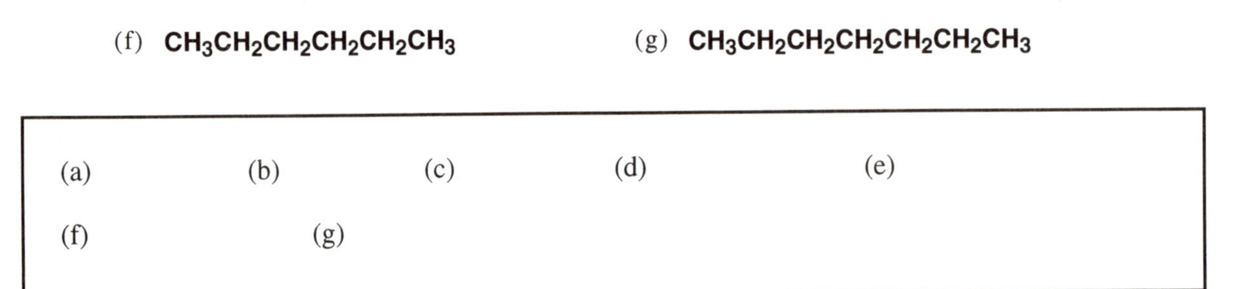

Answer

(a) methane, (b) ethane, (c) propane, (d) butane, (e) pentane, (f) hexane, (g) heptane.

The names for longer alkanes are as follows:

8C octane, 9C nonane, 10C decane, 11C undecane, 12C dodecane, 13C tridecane, 14C tetradecane etc.

4.2 Single-branded alkanes

The above alkanes are all straight-chain alkanes and are easy to name, but what happens if you find yourself faced with a branched alkane? It is best to follow a set procedure and we can demonstrate this with the following example.

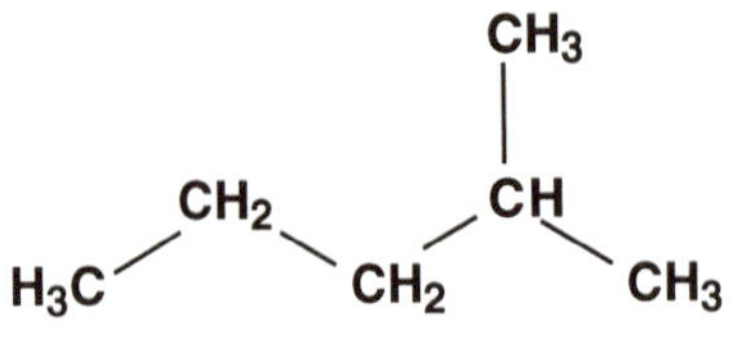

- Stage 1. Identify the longest chain of carbons, i.e.

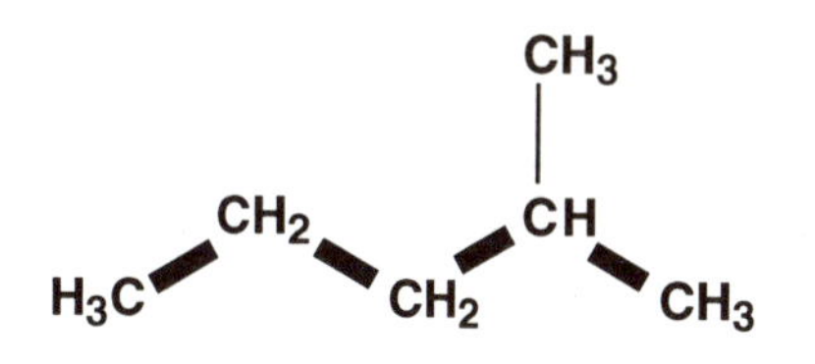

- Stage 2. Number the carbons in the longest chain. Start the numbering from the end nearest the branch point.

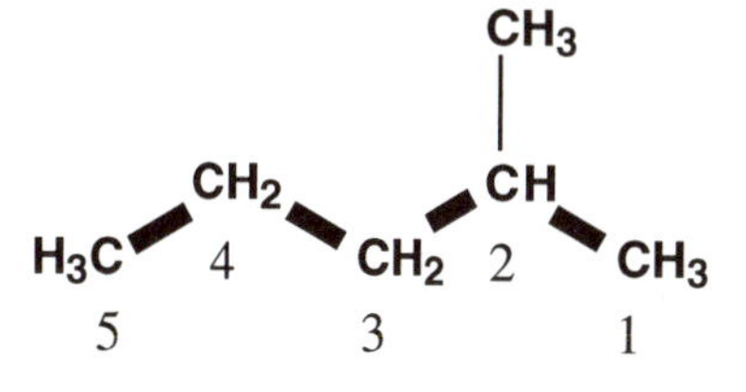

- Stage 3. Identify the name of the longest chain. This has five carbons and is therefore called a **pentane.**
- Stage 4. Identify the carbon with the extra group (number 2 in the example).
- Stage 5. Identify and name the extra group. In this example it is CH_3. Extra groups (or substituents) are named slightly differently from the main chain and are referred to as alkyl groups rather than alkanes. Therefore, CH_3 is called meth**yl** and not meth**ane**.
- Stage 6. You name the structure by first identifying the substituent and its position in the chain. The structure in the example is therefore called 2-methylpentane.

Try this procedure for yourself by naming the following compounds

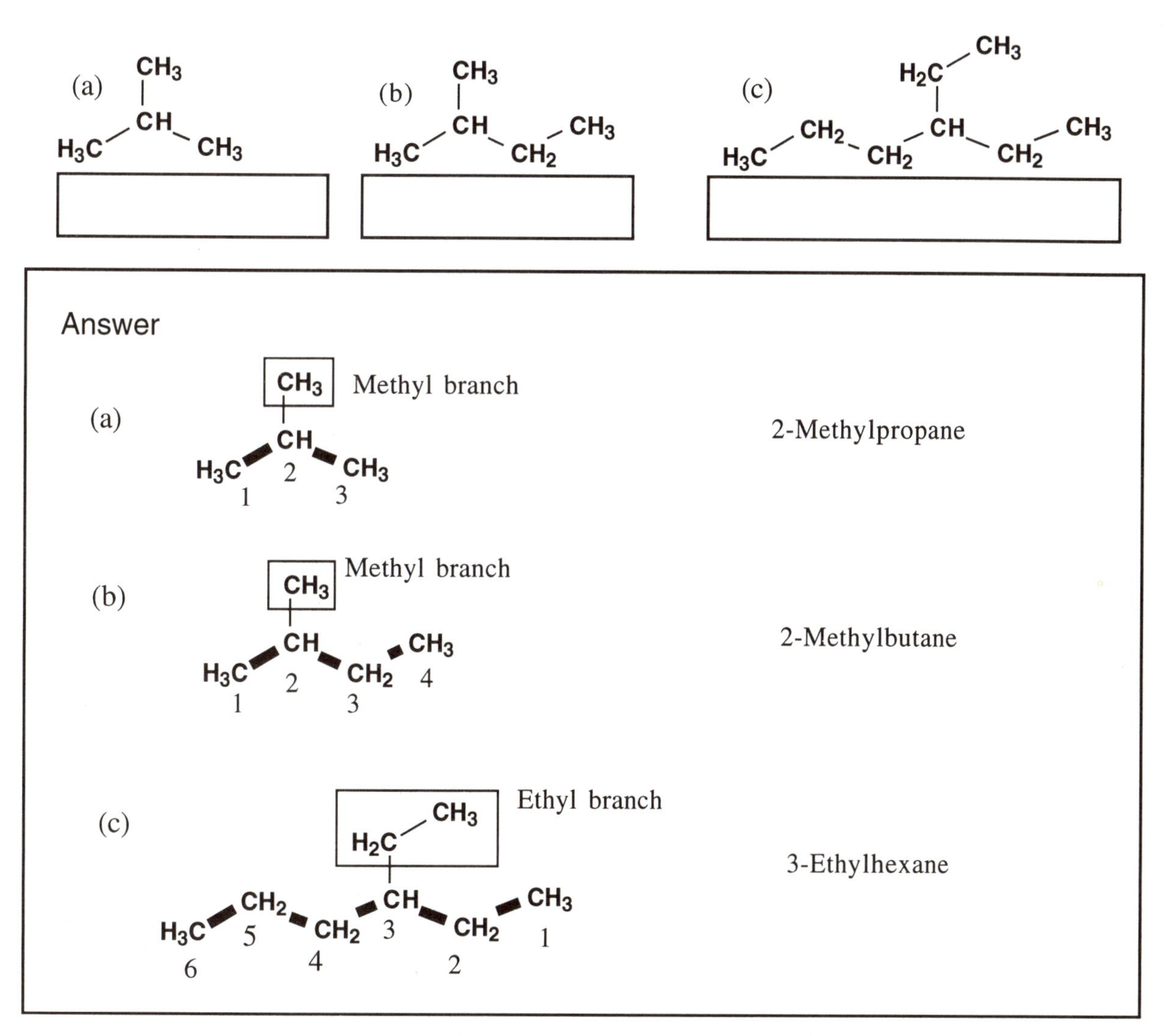

Remember that it is important to identify the longest chain, so beware of examples like the following. These have been set by cunning examiners to see whether students are awake. Try naming them yourself.

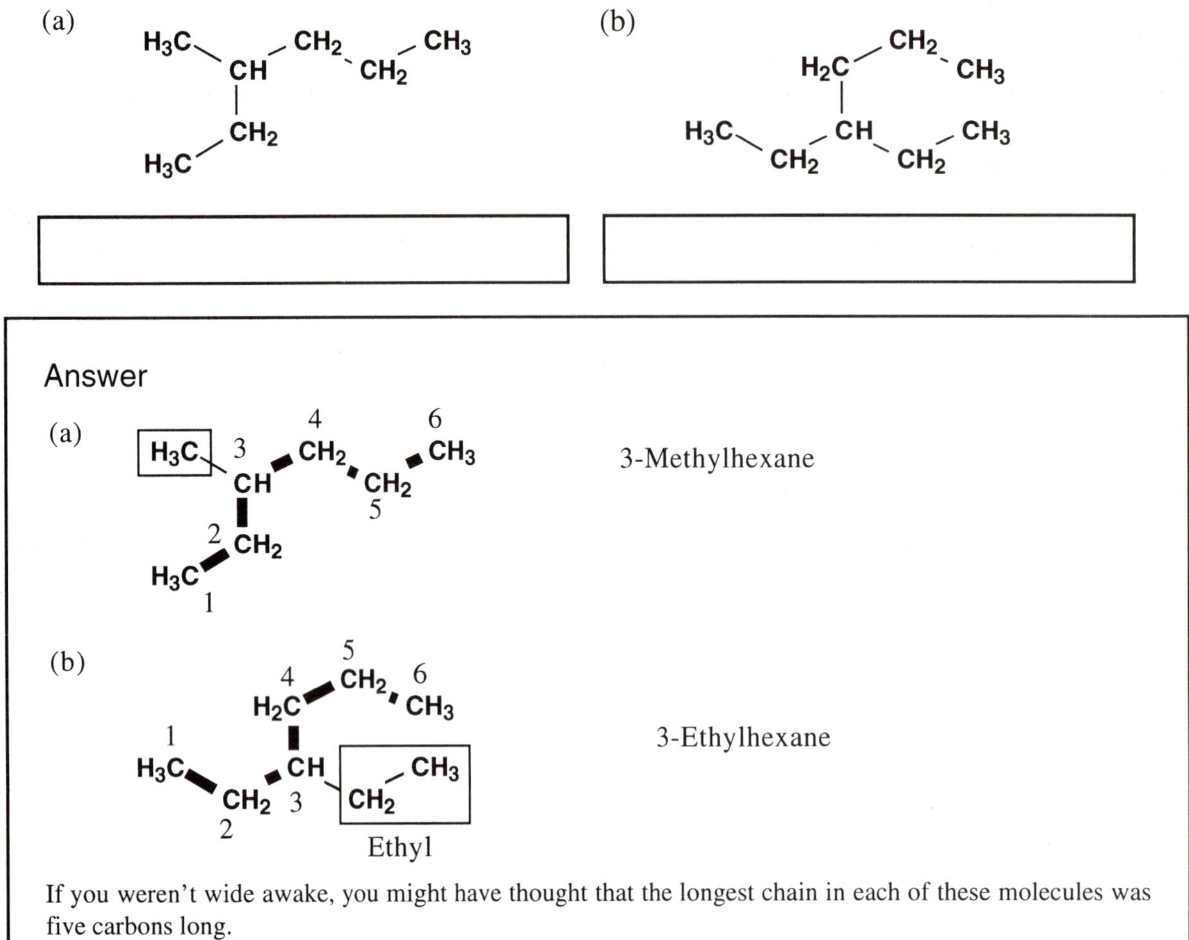

If you weren't wide awake, you might have thought that the longest chain in each of these molecules was five carbons long.

Name the following structures.

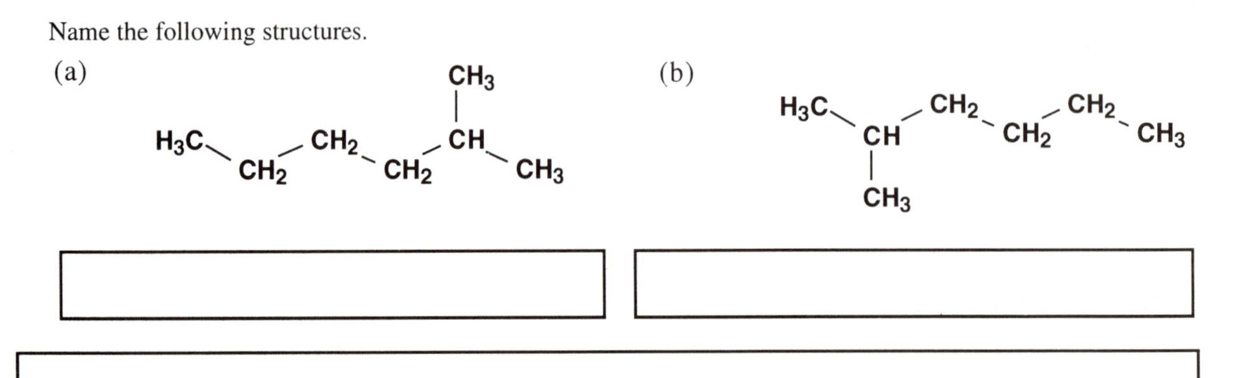

Answer

These are in fact identical molecules. They have been written differently, but that should not make any difference in nomenclature. If you followed the rules correctly, you should have got the same name for both.

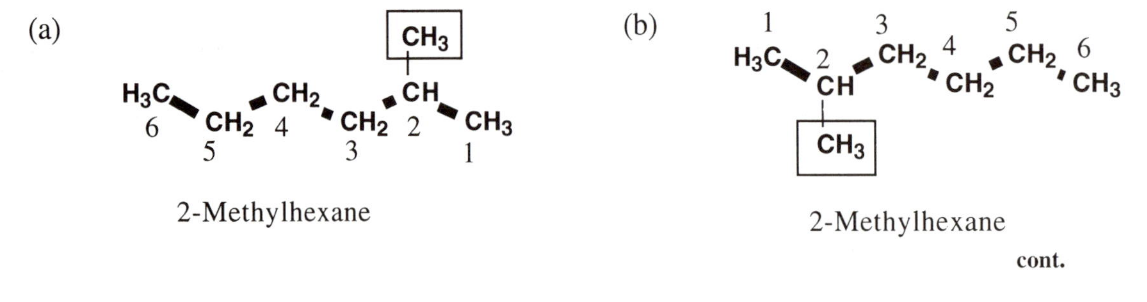

cont.

Remember that the numbering rule states that you must start numbering from the end nearest the branch point.

4.3 Multi-branched alkanes

If you have more than one substituent present in the structure then the substituents are named in alphabetical order, numbering again from the end of the chain nearest the substituents. Name the following molecule.

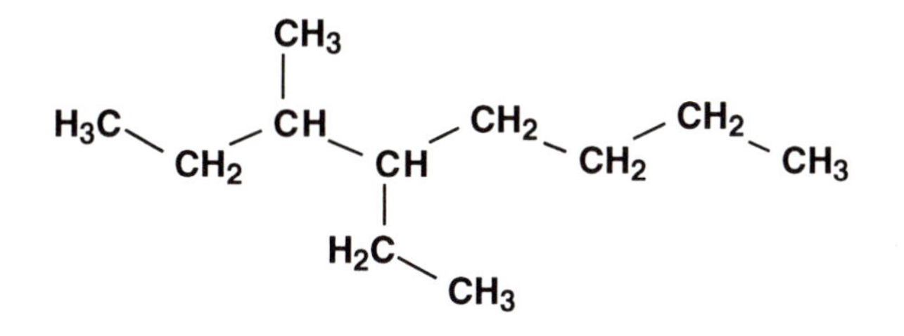

Answer

Octane

Methyl

1 2 3 4 5 6 7 8

Ethyl

4-Ethyl-3-methyloctane

Name the following structures.

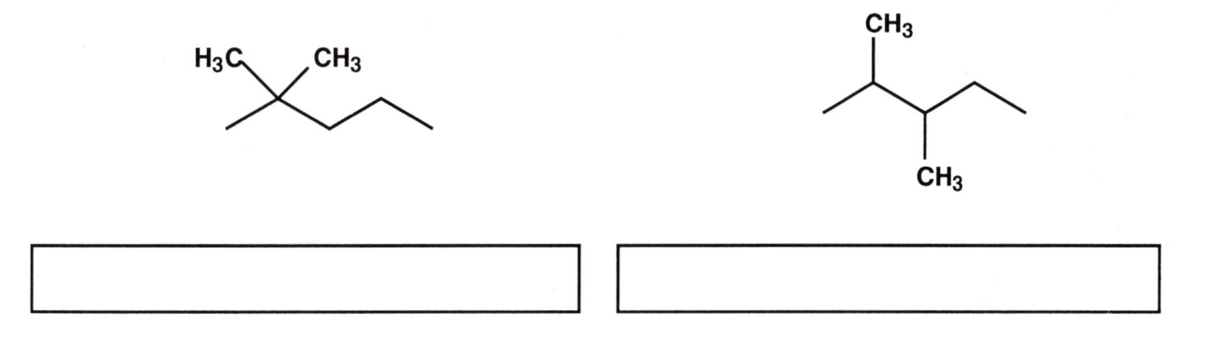

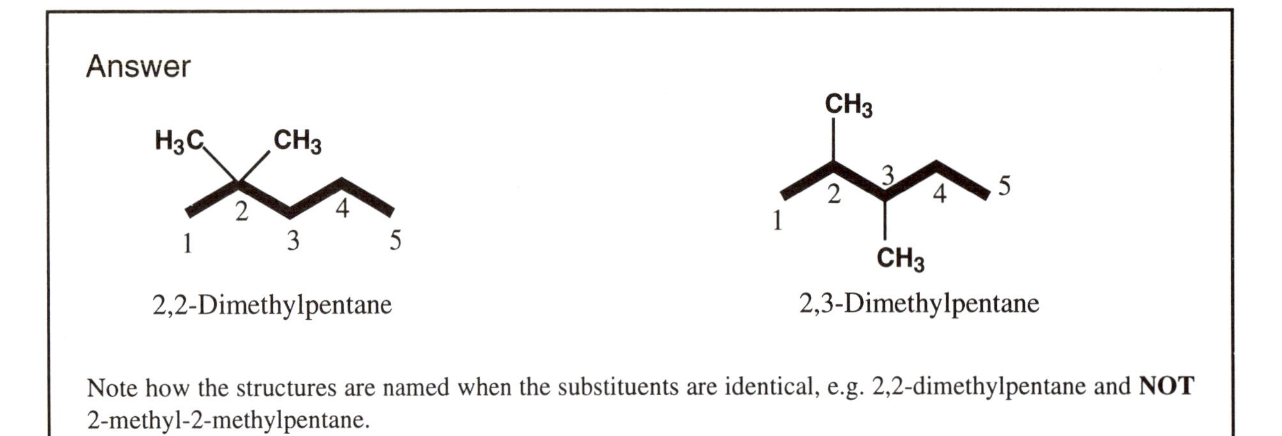

Note how the structures are named when the substituents are identical, e.g. 2,2-dimethylpentane and **NOT** 2-methyl-2-methylpentane.

The prefixes di-, tri-, tetra- etc. are used for identical substituents, but the order in which you write the substituents is still dependent on the alphabetical order of the substituents themselves (i.e. ignore the di-, tri-, tetra- etc.).

Name the following structure.

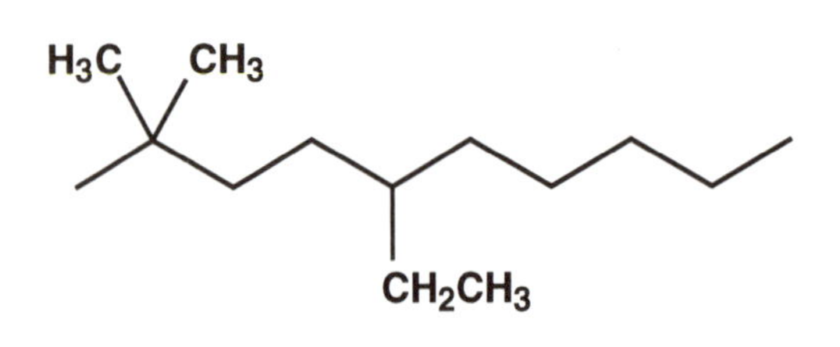

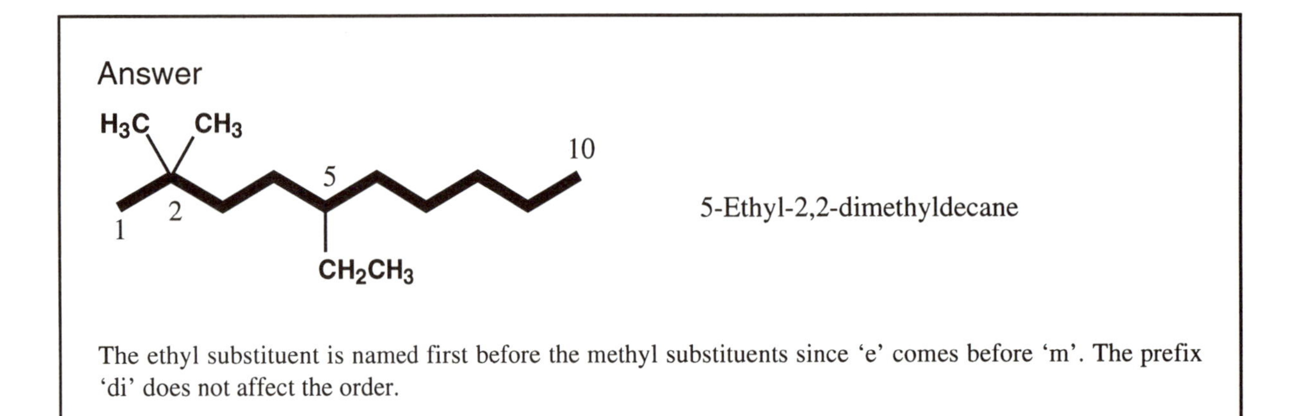

The ethyl substituent is named first before the methyl substituents since 'e' comes before 'm'. The prefix 'di' does not affect the order.

In some structures, it is difficult to decide which end of the chain you should number from. For example, two different substituents might be placed at equal distances from either end of the chain. If that is the case, the group with alphabetical priority should be given the lowest numbering. Try the following structure.

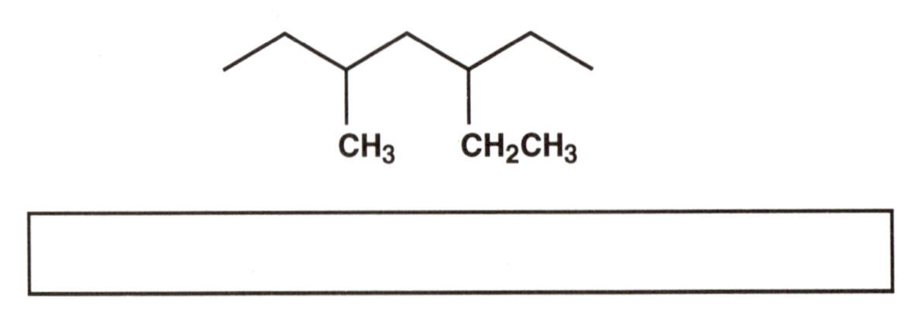

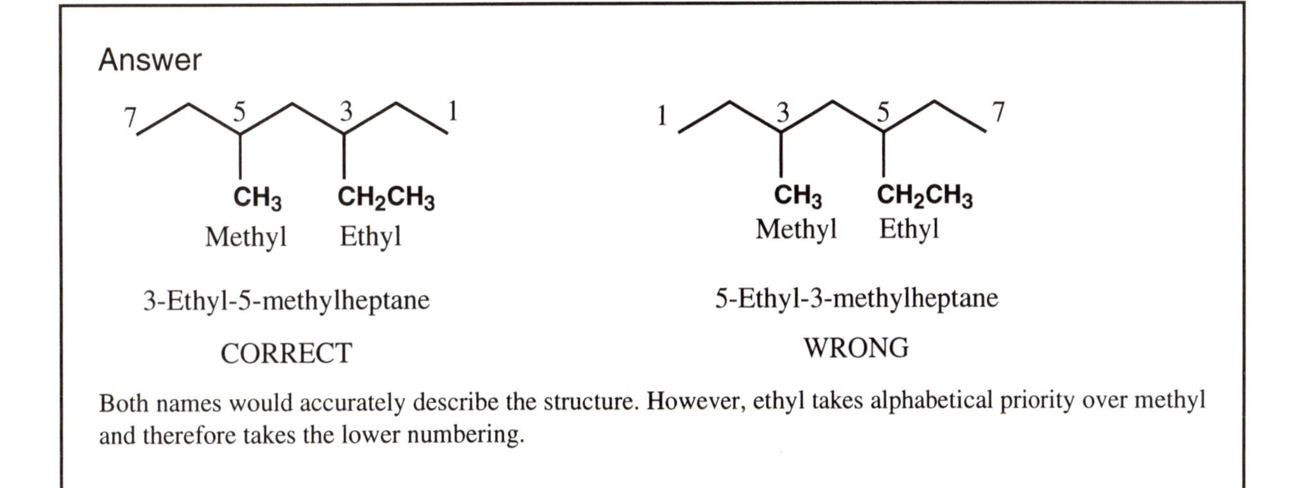

3-Ethyl-5-methylheptane

CORRECT

5-Ethyl-3-methylheptane

WRONG

Both names would accurately describe the structure. However, ethyl takes alphabetical priority over methyl and therefore takes the lower numbering.

Another rule states that the numbering chosen should be such that adding the numbers together gives the smallest total possible. This may take precedence over the previous rule.

Use this rule to name the following structure.

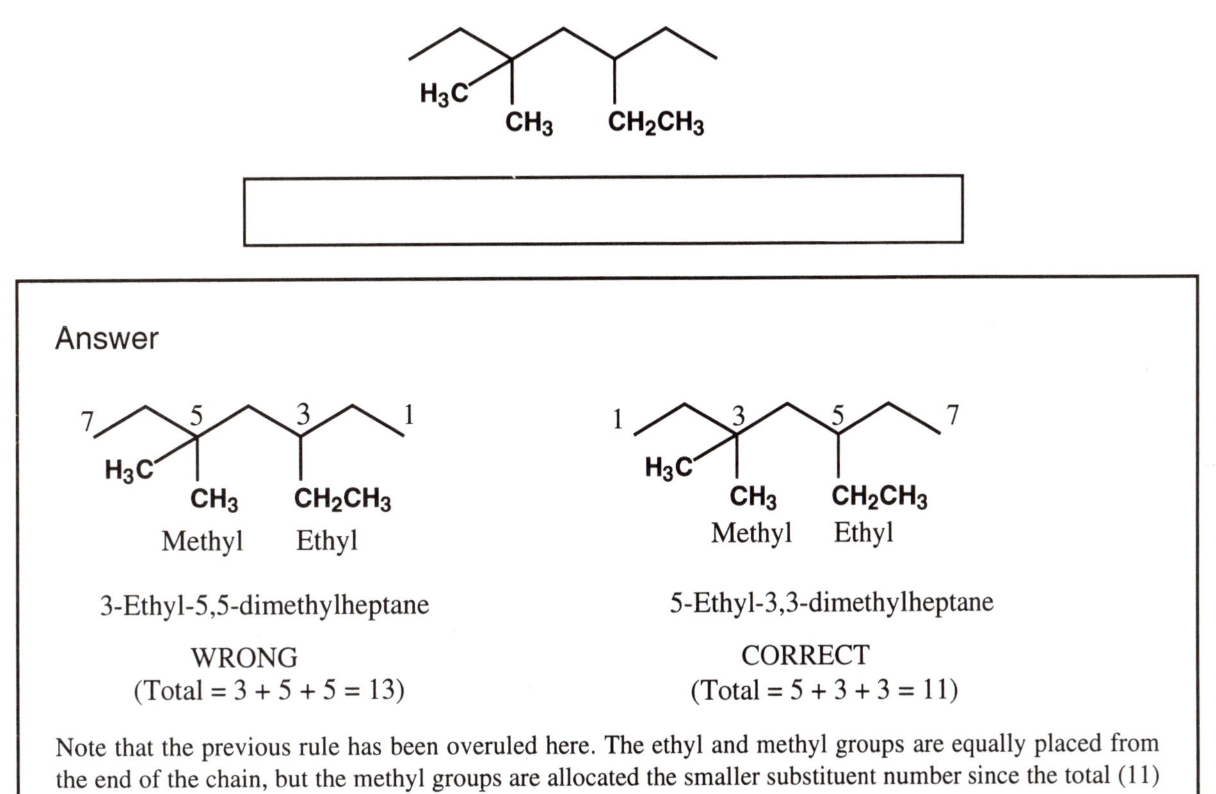

3-Ethyl-5,5-dimethylheptane

WRONG
(Total = 3 + 5 + 5 = 13)

5-Ethyl-3,3-dimethylheptane

CORRECT
(Total = 5 + 3 + 3 = 11)

Note that the previous rule has been overuled here. The ethyl and methyl groups are equally placed from the end of the chain, but the methyl groups are allocated the smaller substituent number since the total (11) is smaller.

Summary

- The naming of alkanes depends on the longest carbon chain in the molecule.
- Alkane chains are named as follows: 1C methane, 2C ethane, 3C propane, 4C butane, 5C pentane, 6C hexane, 7C heptane, 8C octane, 9C nonane, 10C decane etc.
- If the alkane is branched, the numbering starts from the end of the chain nearest the substituent.
- Substituents are named in alphabetical order.
- Substituents are known as alkyl groups, i.e. methyl, ethyl, propyl, butyl, pentyl, hexyl etc.

SECTION 5

Nomenclature—Functional Groups

Nomenclature—Functional Groups

5.1 The rules for functional groups

We now extend the rules of nomenclature to molecules having a functional group. The main rules for such compounds are as follows:

- the presence of a functional group is indicated by replacing -ane for the parent alkane structure with the following suffixes:

Functional Group	Suffix
Alkene	-ene
Alkyne	-yne
Alcohol	-anol
Aldehyde	-anal
Ketone	-anone
Carboxylic acid	-anoic acid
Acid chloride	-anoyl chloride
Amine	-anamine

- the main chain must include the functional group when naming the structure;
- the numbering must start from the end of the main chain nearest the functional group;
- the suffix and position of the functional group is given in the name.

Example

Name the following structure.

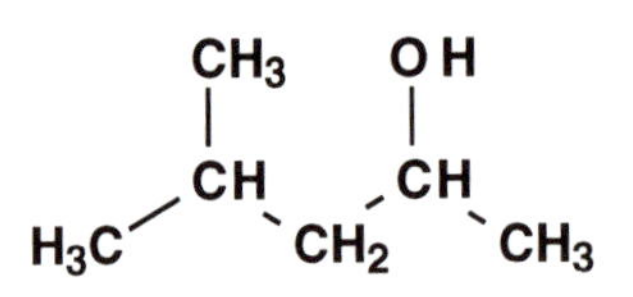

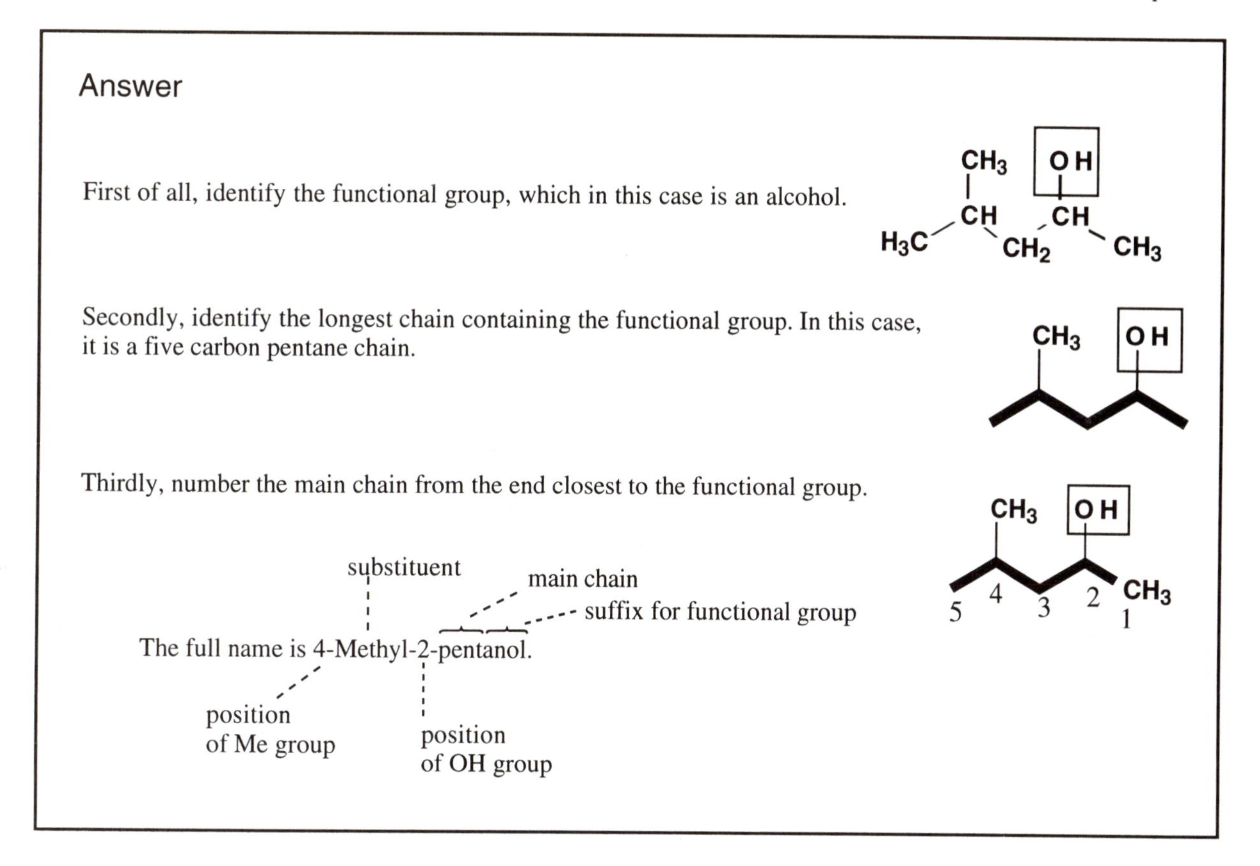

Answer

First of all, identify the functional group, which in this case is an alcohol.

Secondly, identify the longest chain containing the functional group. In this case, it is a five carbon pentane chain.

Thirdly, number the main chain from the end closest to the functional group.

The full name is 4-Methyl-2-pentanol.

5.2 Nomenclature of alcohols

We have already discussed the naming of an alcohol in the above example. Name the following structures.

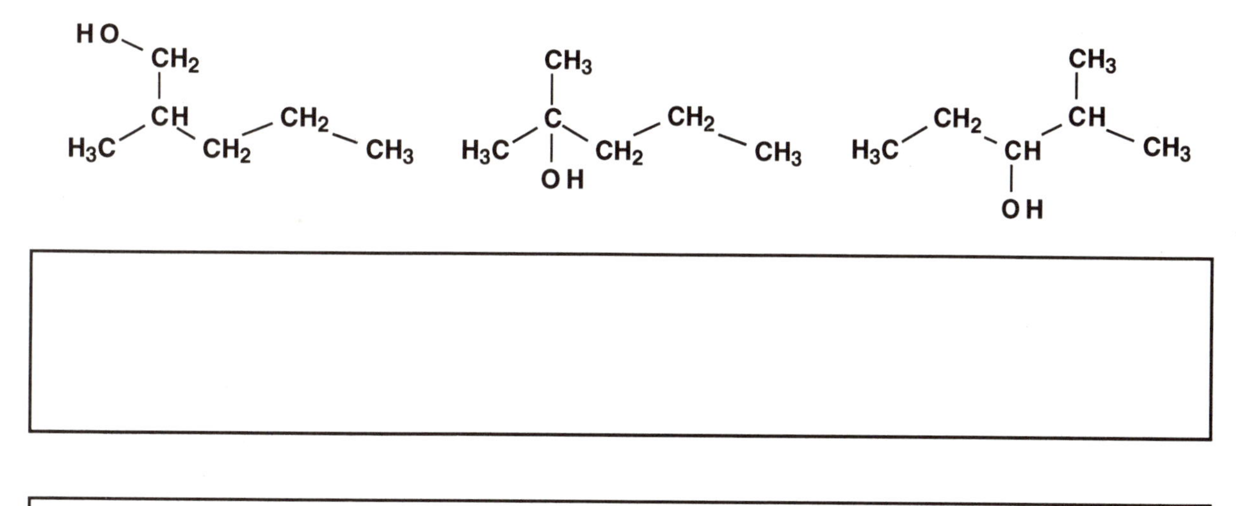

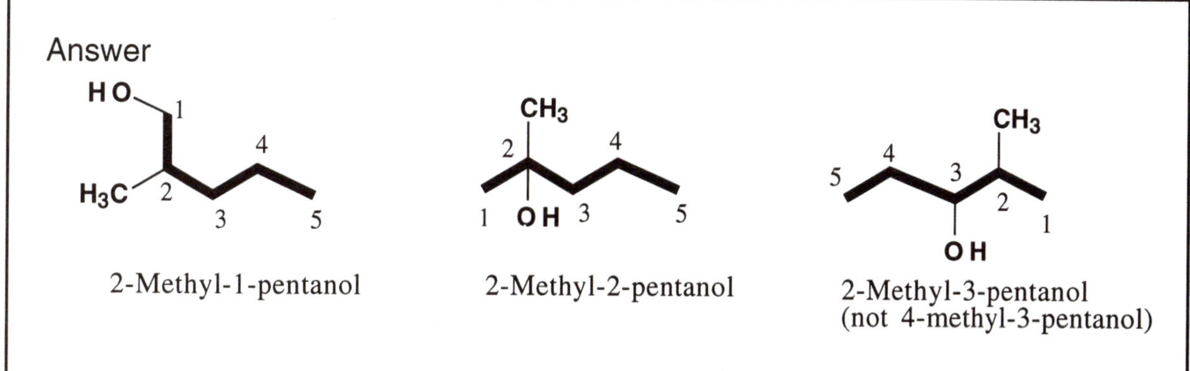

Answer

2-Methyl-1-pentanol

2-Methyl-2-pentanol

2-Methyl-3-pentanol
(not 4-methyl-3-pentanol)

5.3 Nomenclature of compounds containing a carbonyl group (C=O)

The following examples are all ketones. Name them, remembering that the suffix is -anone.

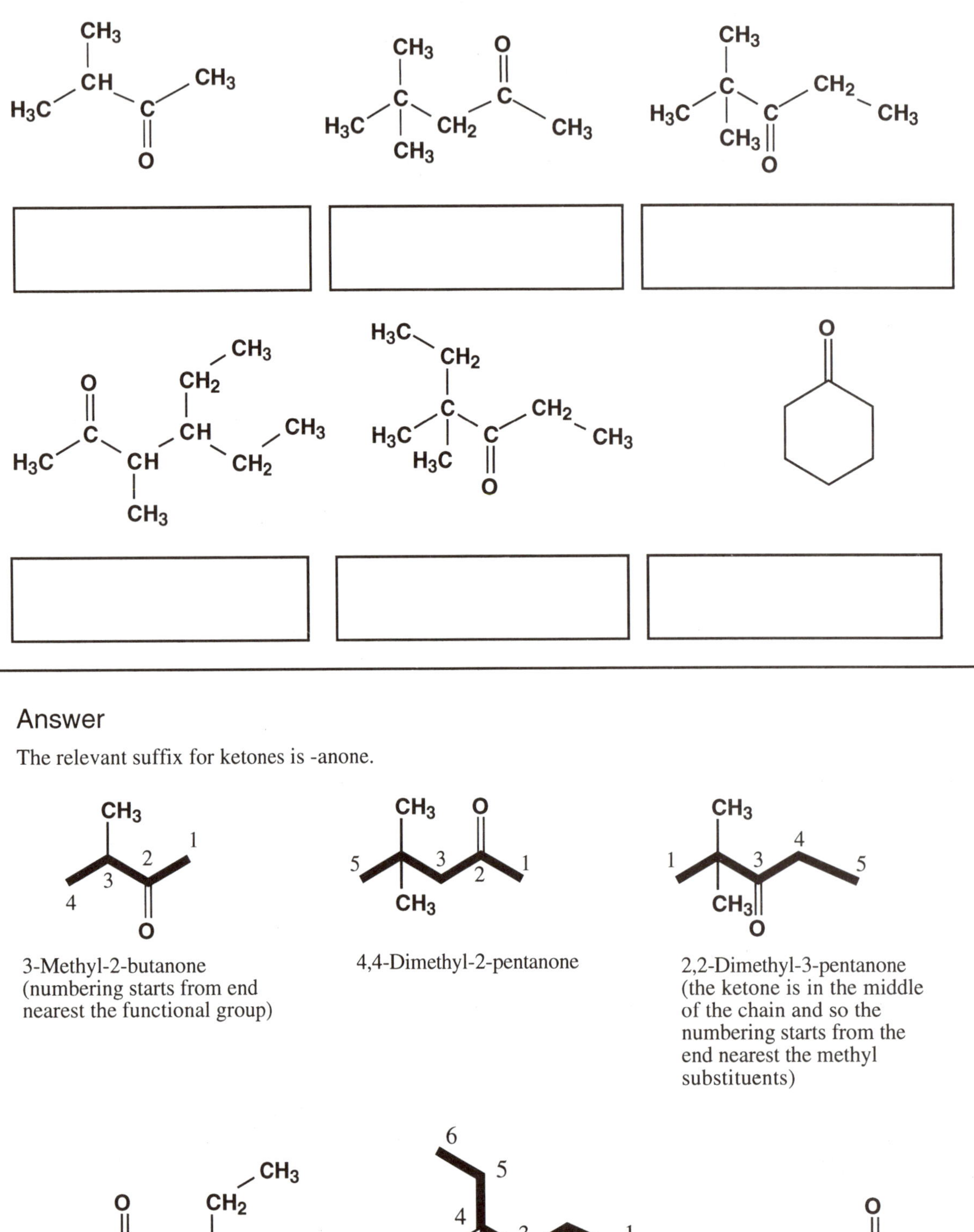

Answer

The relevant suffix for ketones is -anone.

3-Methyl-2-butanone (numbering starts from end nearest the functional group)

4,4-Dimethyl-2-pentanone

2,2-Dimethyl-3-pentanone (the ketone is in the middle of the chain and so the numbering starts from the end nearest the methyl substituents)

4-Ethyl-3-methyl-2-hexanone (not 3-methyl-4-ethyl-2-hexanone, i.e. alphabetical order of substituents)

4,4-Dimethyl-3-hexanone (not 3,3-dimethyl-4-hexanone, Numbering from end closest to functional group)

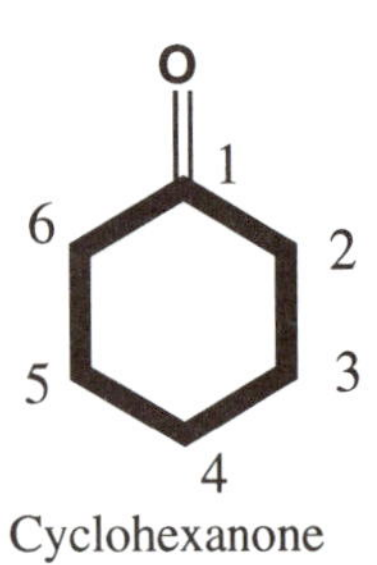

Cyclohexanone

The following molecule was in the previous example and was named 3-methyl-2-butanone. In fact, the name can be simplified to 3-methylbutanone. Why?

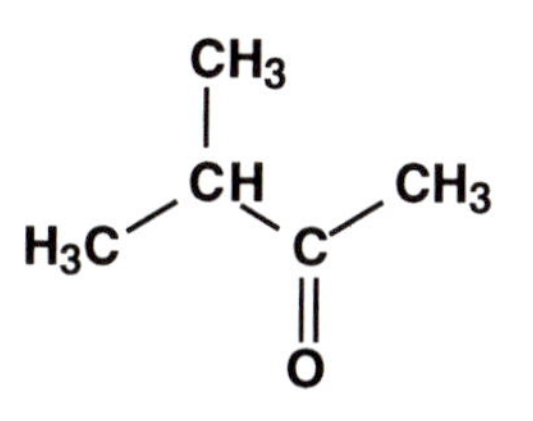

Answer

There is only one possible place for the ketone functional group in this molecule. If the carbonyl C=O group was at the end of the chain, it would be an aldehyde and not a ketone. Therefore, the number 2 is not necessary in identifying the structure.

Numbering is not necessary in locating certain functional groups. For which functional groups is this true and why? (The previous answer gives you a hint.)

Answer

Any functional group which must be at the end of a chain will always be at position one and therefore it is not necessary to number them. Such functional groups are aldehydes, carboxylic acids and acid chlorides.

Name the following structures.

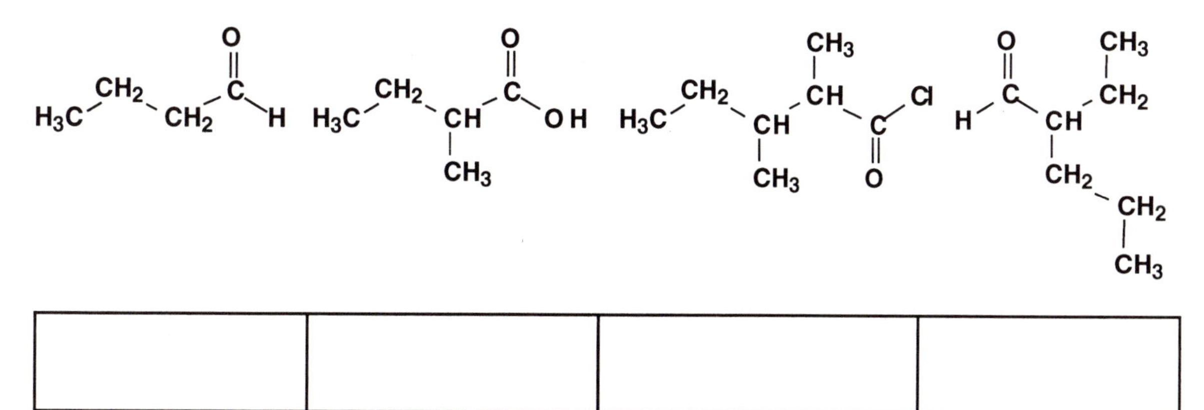

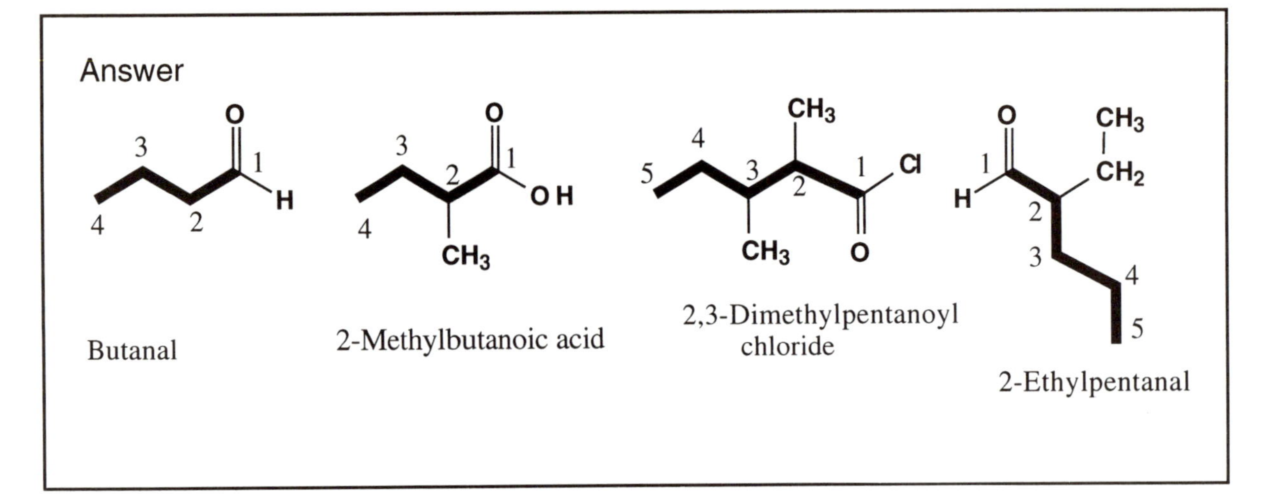

5.4 Nomenclature of alkenes and alkynes

Name the following alkenes and alkynes, remembering that the suffixes are -ene and -yne, respectively.

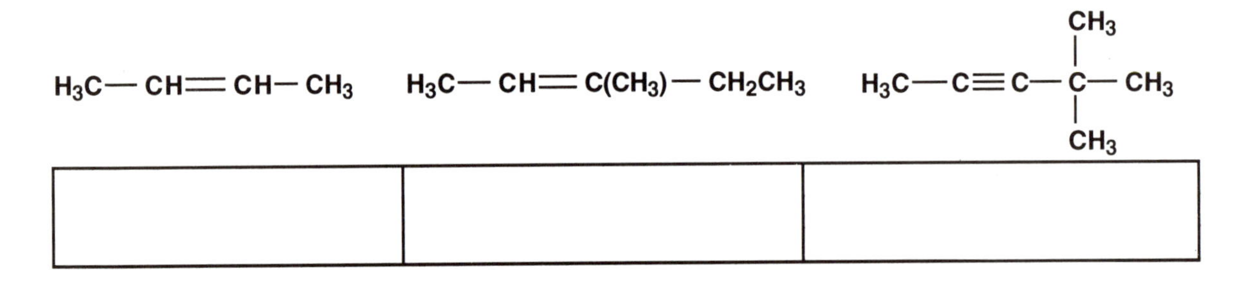

Answer

Notice that with some alkenes it is necessary to say whether they are *cis* or *trans*. This will be covered in Section 11.

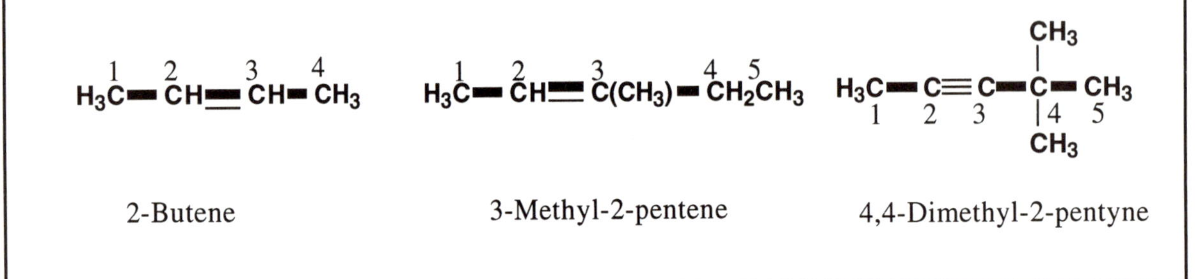

5.5 Nomenclature of esters

It is helpful to realise from the onset that esters are formed from a carboxylic acid and an alcohol. Thus, two different molecules are coming together, reacting and linking up with the loss of water, as shown below.

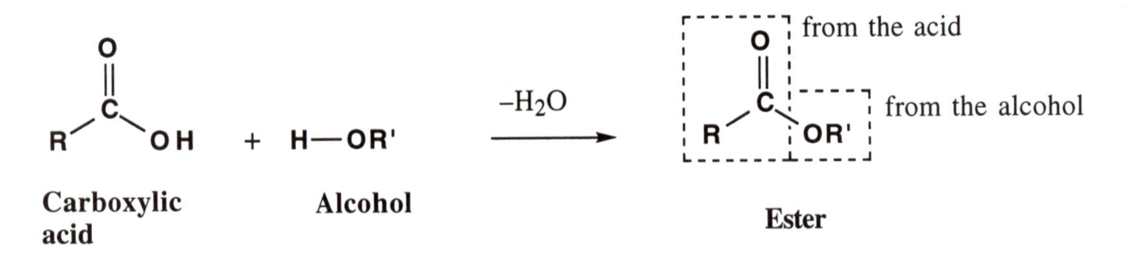

To name the ester:

- Identify the carboxylic acid (alkanoic acid) from which it was derived.
- Change the name to an alkanoate rather than an alkanoic acid.
- Identify the alcohol from which the ester was derived and consider this as an alkyl substituent.
- The name becomes an alkyl alkanoate.

Example

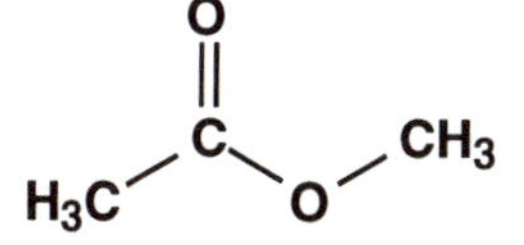

- The above ester would be derived from ethanoic acid and methanol.

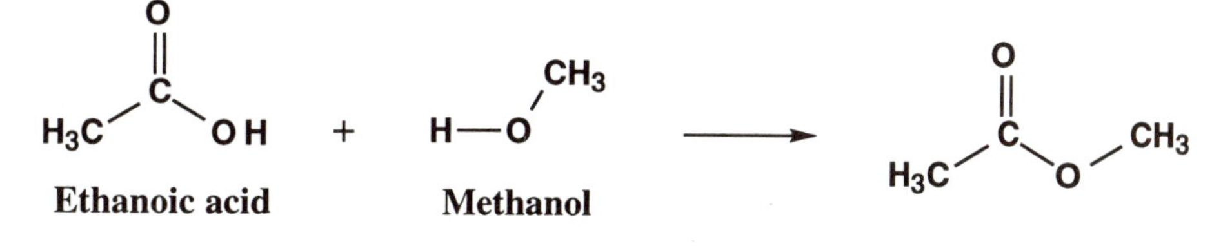

- The ester would be an alkyl ethanoate since it is derived from ethanoic acid.
- The alkyl group comes from methanol and is therefore a methyl group.
- The full name is methyl ethanoate. (Note: there should be a space between both parts of the name.)

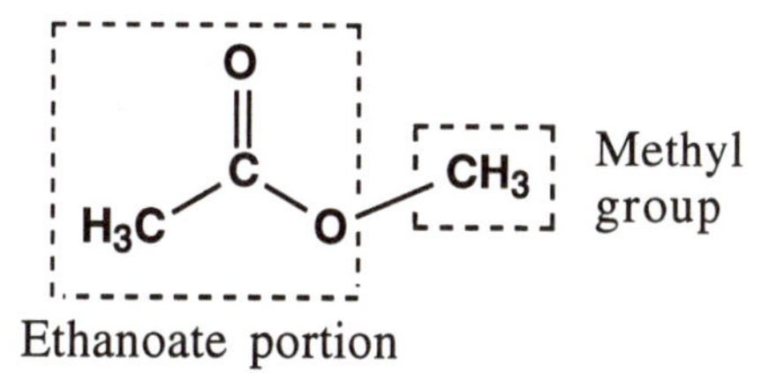

(Note: Ethanoic acid is commonly called acetic acid, so it is also acceptable to call this compound methyl acetate. These alternative names were used before the introduction of the IUPAC rules on nomenclature, and are still commonly used by chemists today.)

Name the following esters.

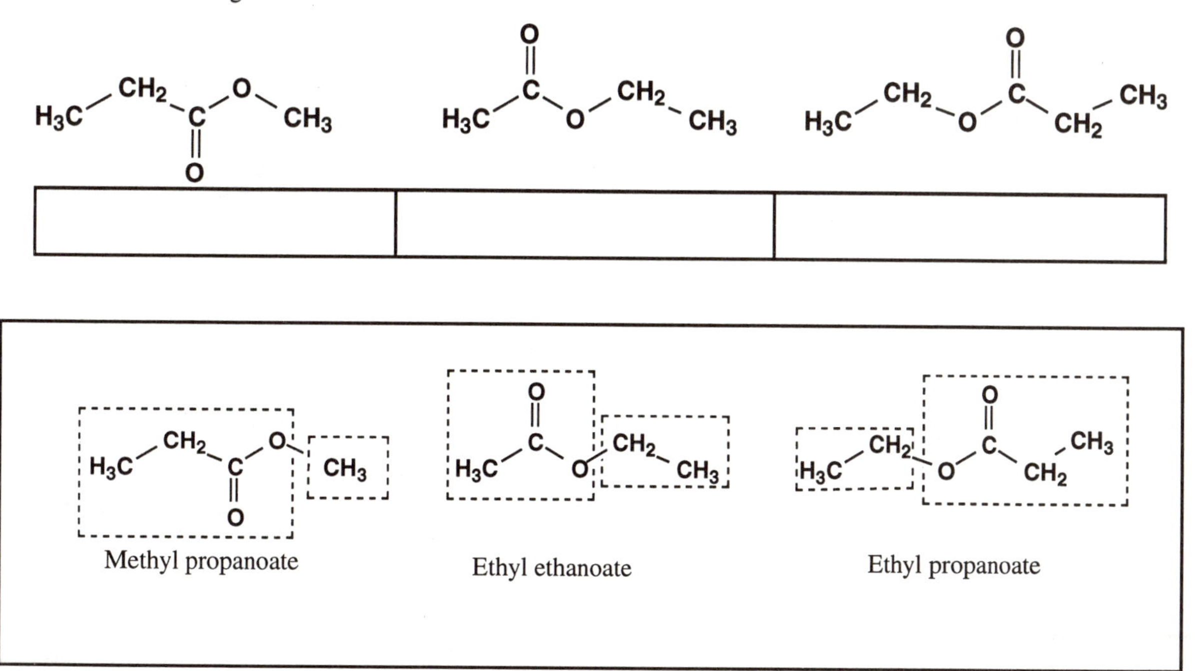

The following is a more complex problem. In this example, you will have to use numbering to show where the substituents are. Once again, identify the carboxylic acid and the alcohol which were used to make the ester before you name the ester itself.

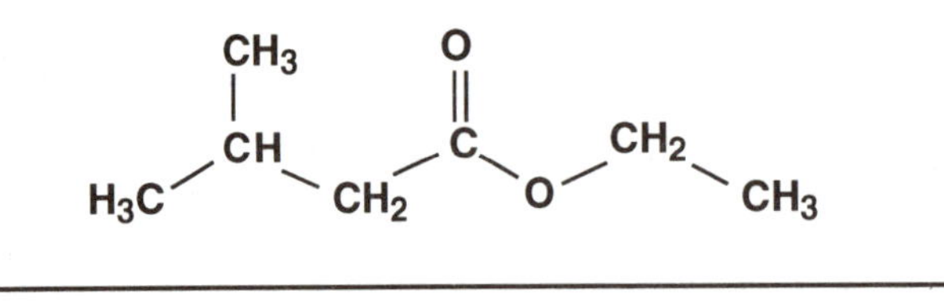

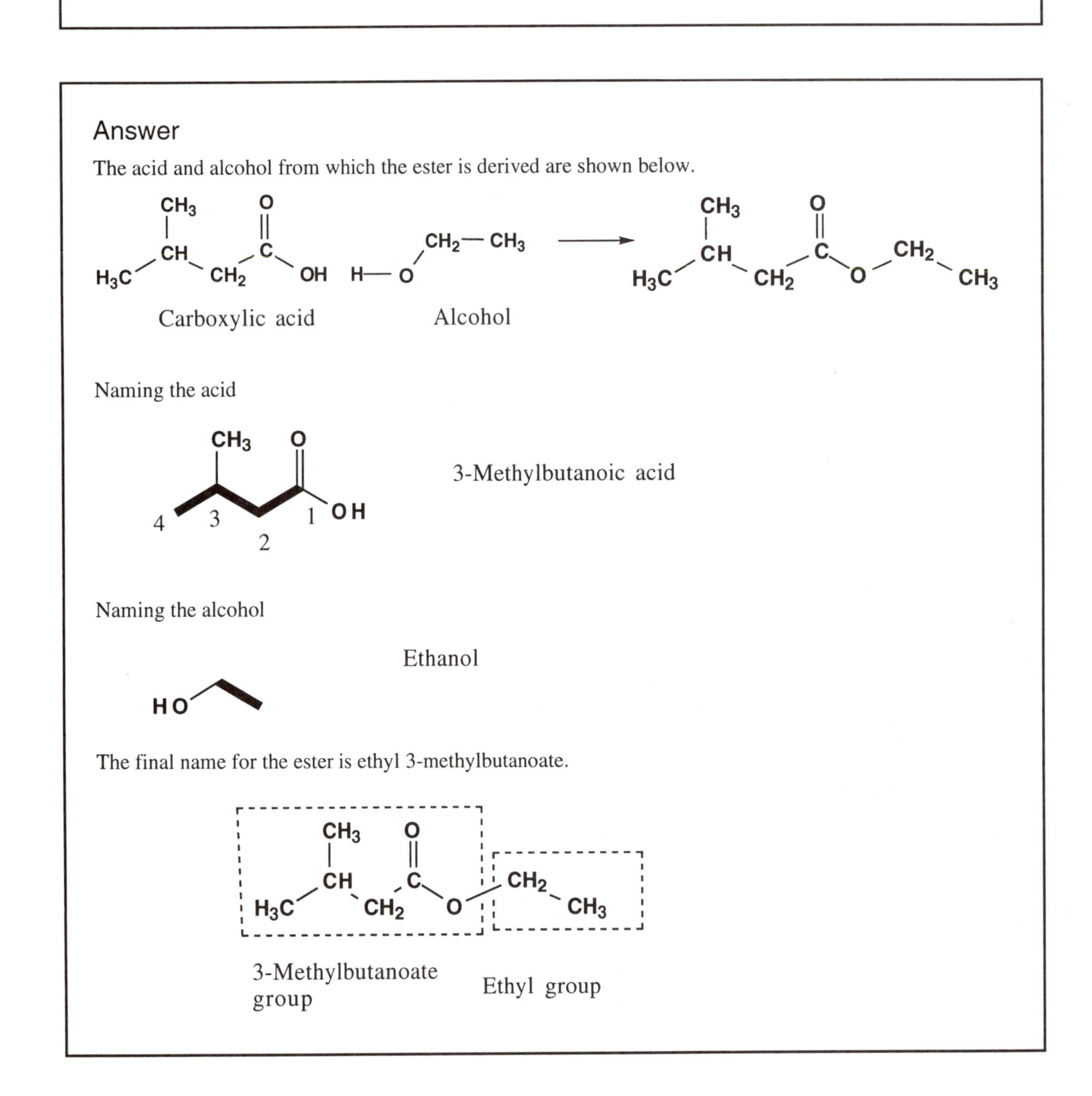

Answer

The acid and alcohol from which the ester is derived are shown below.

Naming the acid

3-Methylbutanoic acid

Naming the alcohol

Ethanol

The final name for the ester is ethyl 3-methylbutanoate.

5.6 Nomenclature of amides

Let us now consider the nomenclature of amides. These are also derivatives of carboxylic acids. This time the carboxylic acid has been linked with an amine rather than an alcohol.

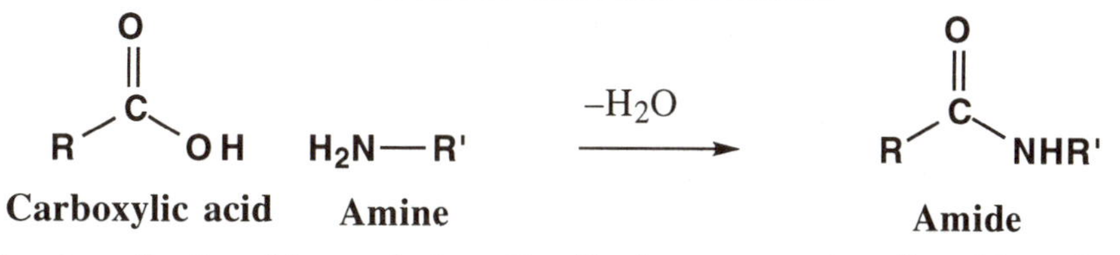

As with esters, the first thing to do is to identify the parent carboxylic acid or alkanoic acid. This is then termed as the alkanamide. Unlike esters, you do not need to name the other component (i.e. the amine). If the amide has alkyl groups then these are considered as alkyl substituents and come at the beginning of the name. The symbol *N* is used to show that the substituents are on the nitrogen and not some other part of the alkanamide skeleton.

Example

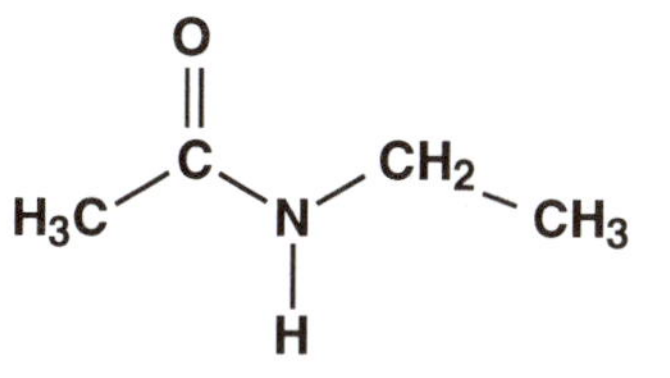

This amide is derived from the following carboxylic acid and amine:

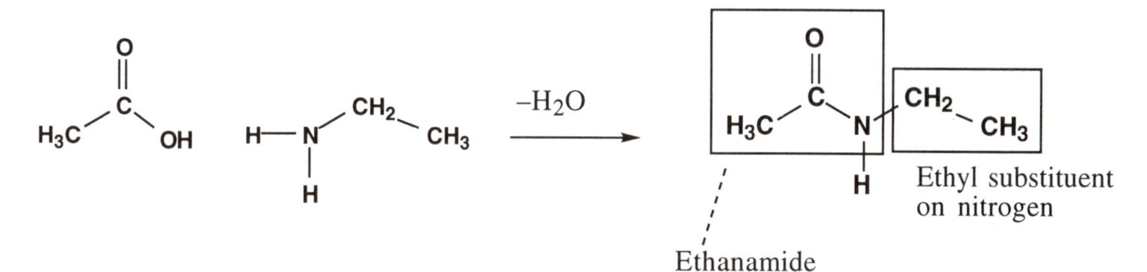

The name for the amide is *N*-ethylethanamide.

Name the following amides.

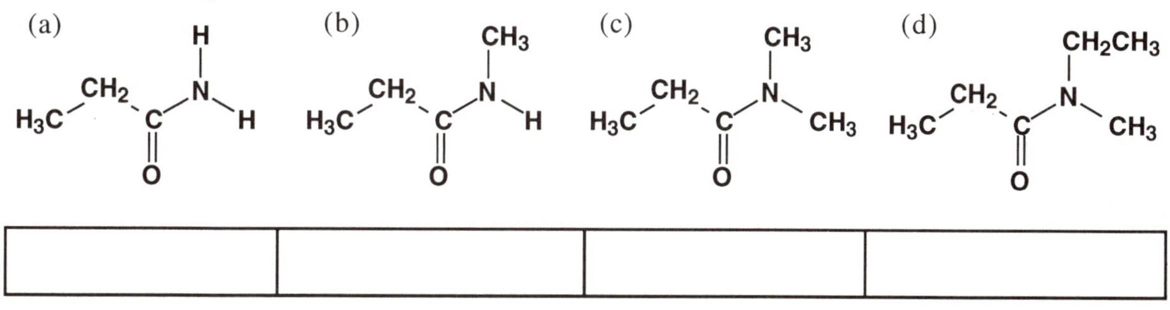

Answer

All these examples are derived from propanoic acid and are therefore propanamides.

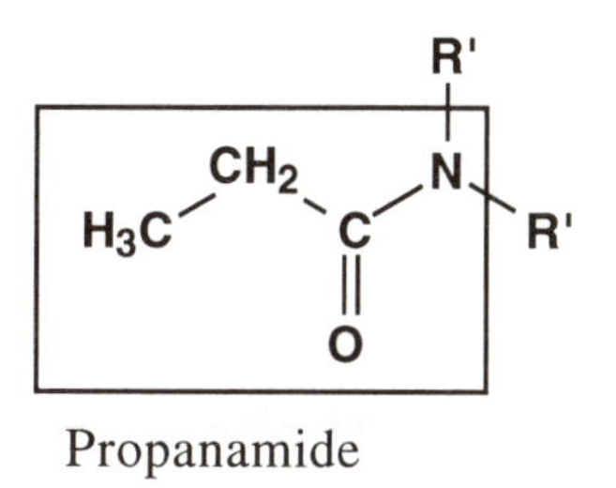

cont.

Any alkyl group on the nitrogen is a substituent and mentioned at the start of the name. The answers are as follows:

(a) propanamide, (b) *N*-methylpropanamide, (c) *N*, *N*-dimethylpropanamide, (d) *N*-ethyl-*N*-methylpropanamide.

5.7 Nomenclature of ethers and alkyl halides (haloalkanes)

The nomenclature for these compounds is slightly different from previous examples in that the main part (or root) of the name is an alkane. The halogen is considered to be a substituent and is numbered accordingly.

Examples

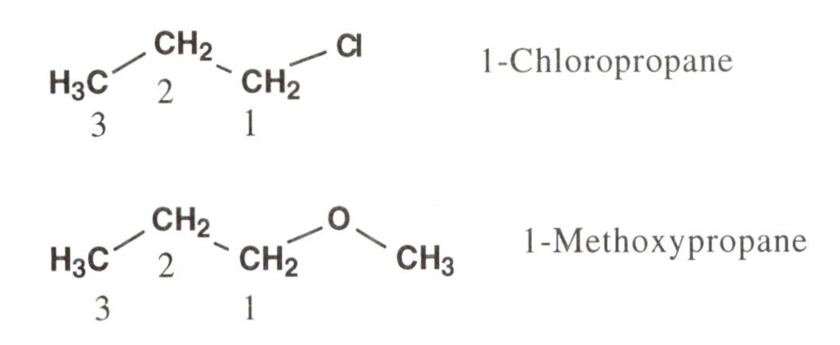

Note that ethers have alkyl groups on both sides of the oxygen. The larger alkyl group is the 'root' alkane. The smaller alkyl group along with the oxygen is known as the alkoxy group and placed at the beginning of the name in the same way as a halogen.

Give names for the following haloalkanes.

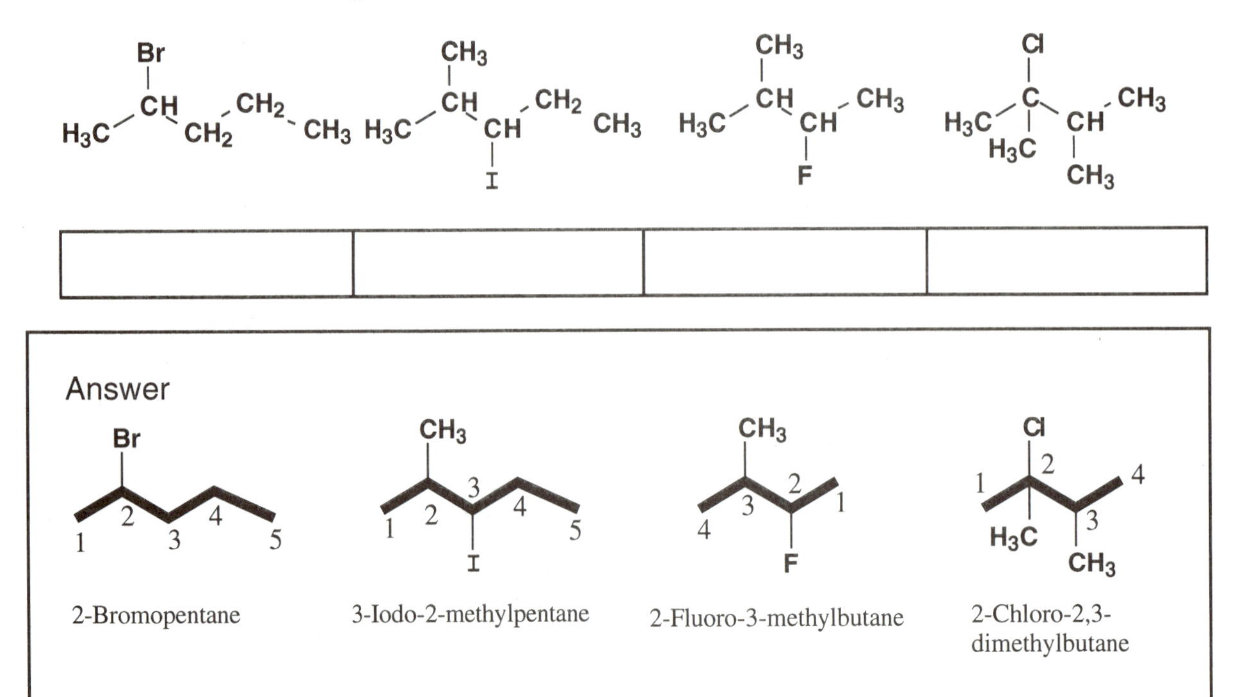

Name the following ethers.

$H_3C-CH_2-CH_2-O-CH_3$ $\quad$ $H_3C-CH_2-O-CH_2-CH_2-CH_3$ $\quad$ $H_3C-CH_2-O-CH_2-CH_3$

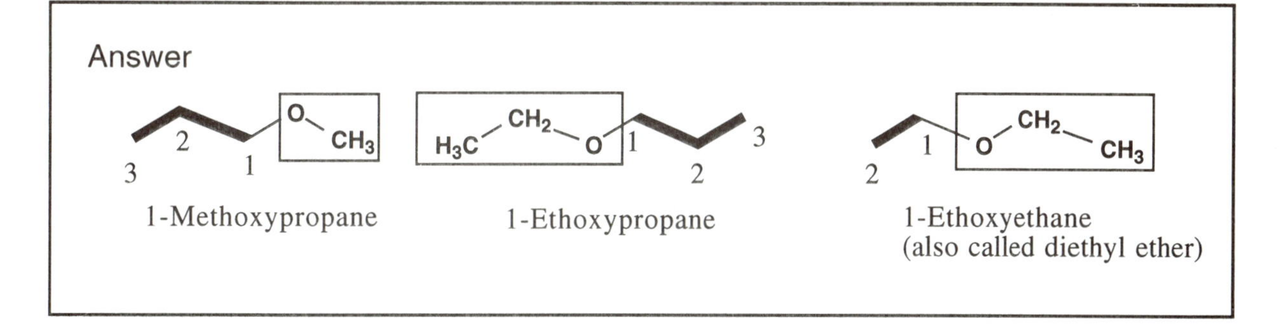

5.8 Nomenclature of amines

The nomenclature for primary amines is similar to alcohols using the suffix -anamine. The suffix -ylamine is used for simple amines as well.

Examples

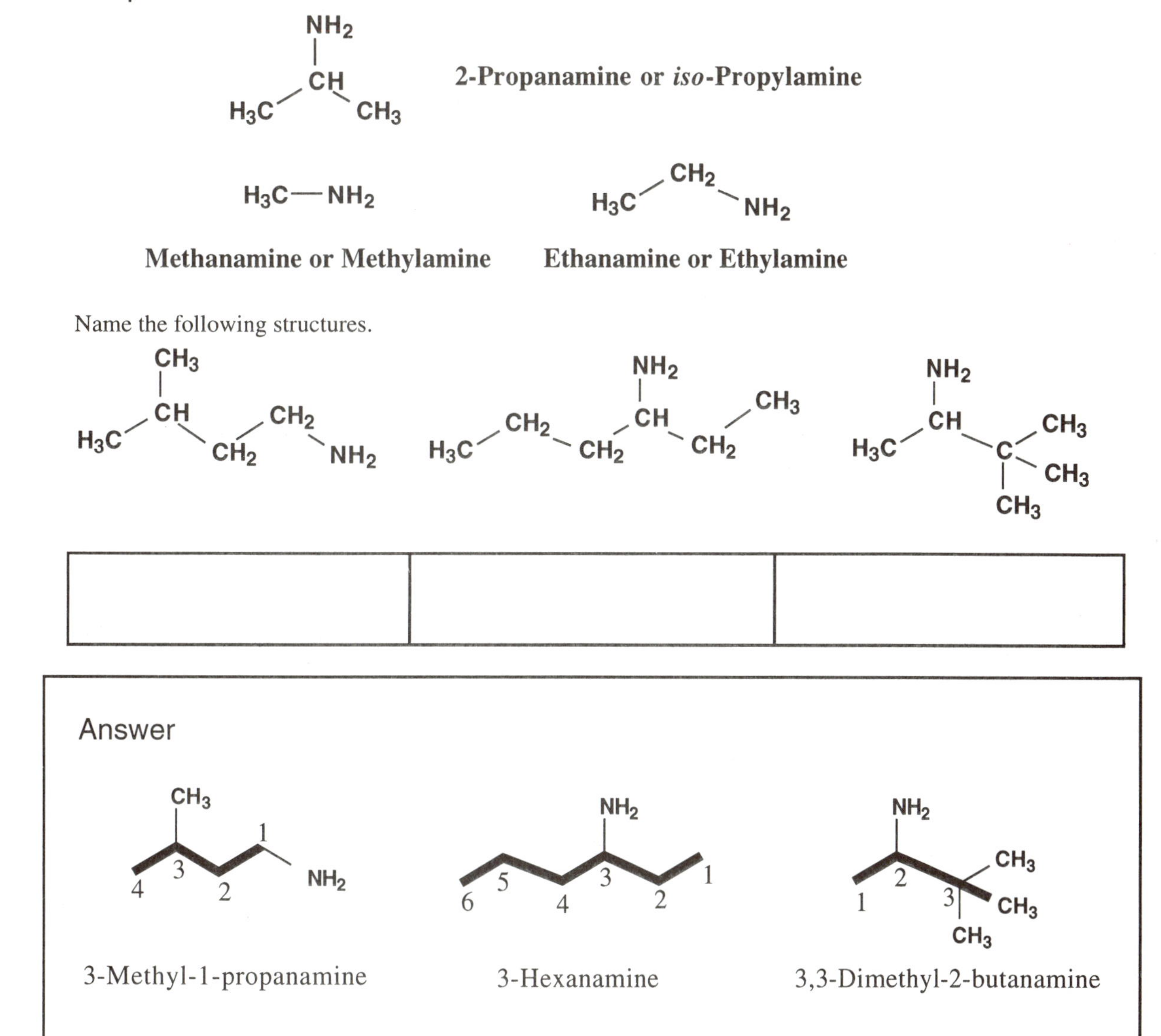

The examples above only have one alkyl group attached to the nitrogen. How would you name amines which have more than one alkyl group attached i.e. secondary and tertiary amines?

For example, how would you name the following?

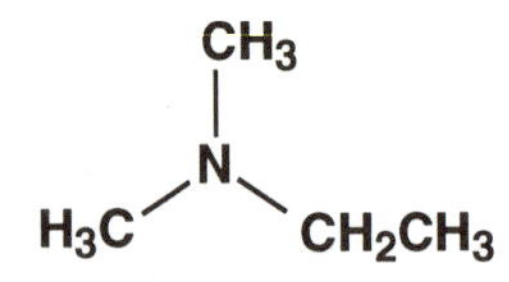

The first thing to do is to identify the longest carbon chain attached to the nitrogen.

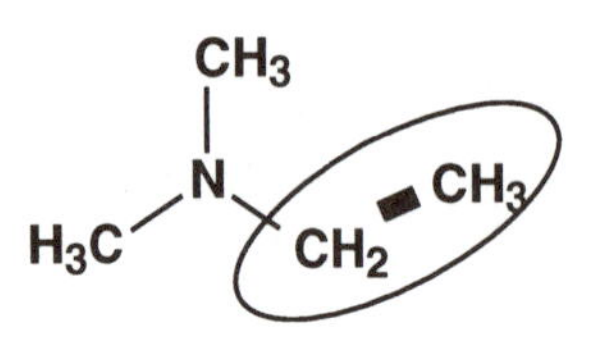

This is an ethane and so this is an ethanamine

The remaining alkyl groups are considered to be substituents on the nitrogen atom and so the full name for the structure would be *N*,*N*-Dimethylethanamine.

Draw the structures for the following compounds:

(a) *N*-Ethyl-*N*-methyl-1-pentanamine,
(b) 2-Ethyl-3-methyl-1-pentanamine.

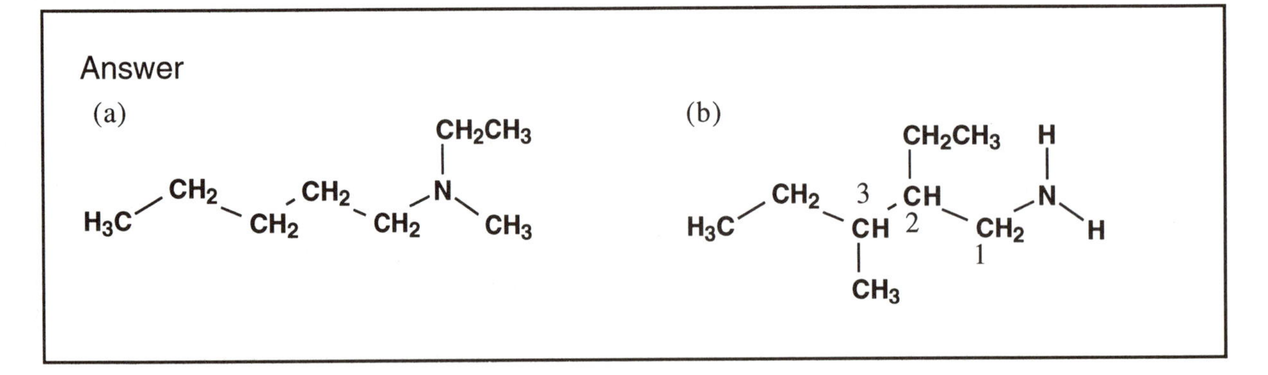

Give the names for the following amines.

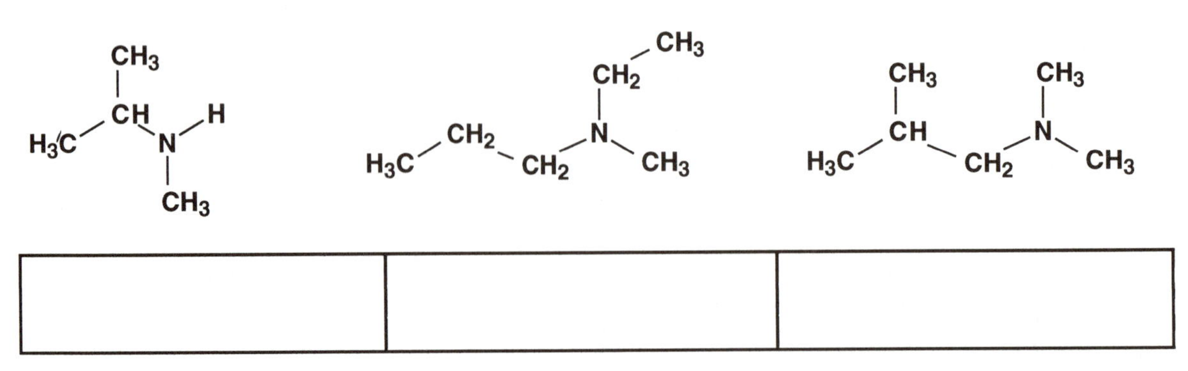

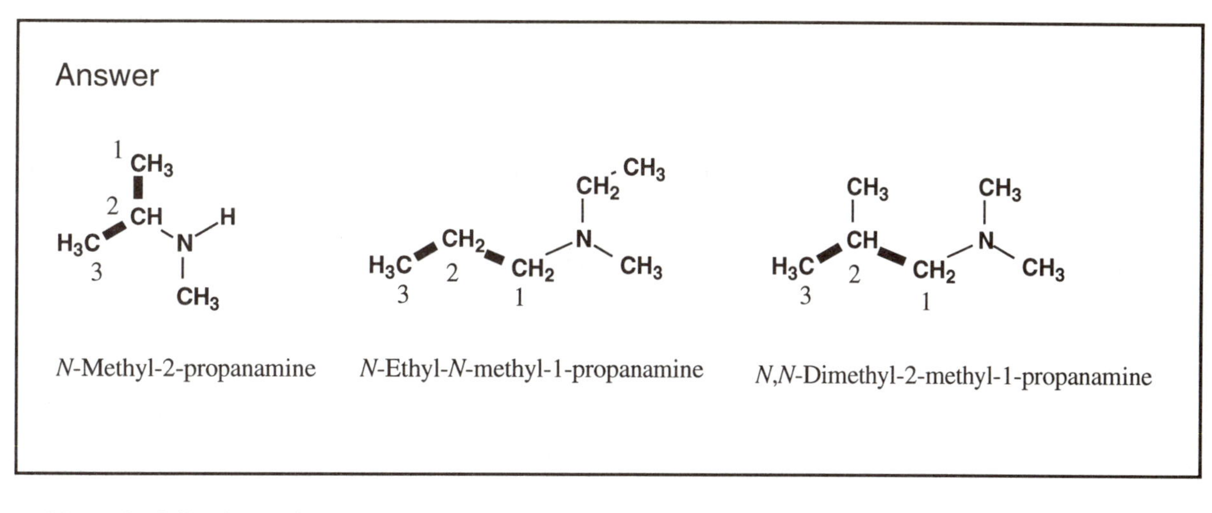

N-Methyl-2-propanamine   *N*-Ethyl-*N*-methyl-1-propanamine   *N*,*N*-Dimethyl-2-methyl-1-propanamine

Name the following amines.

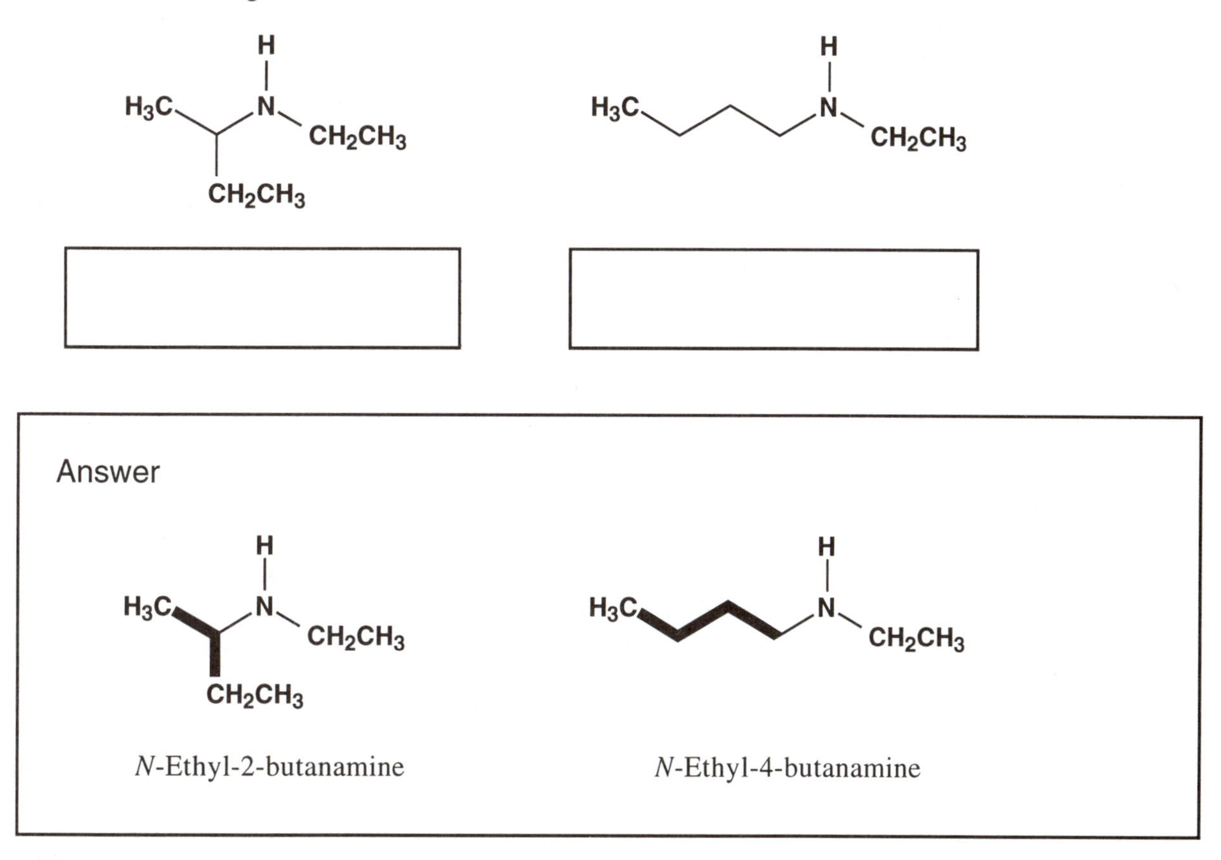

N-Ethyl-2-butanamine   *N*-Ethyl-4-butanamine

It is worth noting that some simple secondary and tertiary amines have common names.

Examples

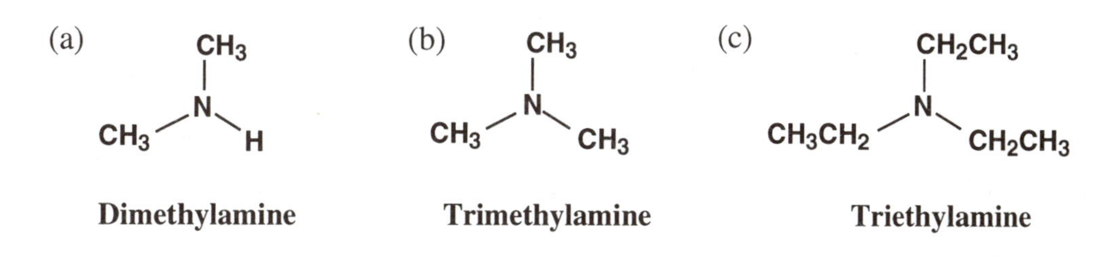

Dimethylamine   **Trimethylamine**   **Triethylamine**

Summary

Compounds containing functional groups are named according to the following rules.

- The main chain must be chosen such that it includes the functional group.
- Numbering of the main chain must start from the end nearest the functional group.
- The suffix -ane is removed from the main chain and replaced with the suffix for the functional group.

Functional Group	Suffix
Alkene	-ene
Alkyne	-yne
Alcohol	-anol
Aldehyde	-anal
Ketone	-anone
Carboxylic ccid	-anoic acid
Acid chloride	-anoyl chloride
Amine	-anamine

- Other substituents are named and ordered in the same way as for alkanes.
- If the functional group is an equal distance from either end of the main chain, the numbering starts from the end of the chain nearest to any substituents.
- The nomenclature *cis* and *trans* may be necessary for alkenes (Section 11).
- The functional group for aldehydes, carboxylic acids and acid chlorides will always be at position 1 and does not need to be numbered.
- Esters are named from the parent carboxylic acid and alcohol. The alkanoic acid is renamed alkanoate and the alkanol is treated as an alkyl substituent. The combined name is alkyl alkanoate. There must be a space between both parts of the name.
- Amides are named from the carboxylic acid from which they are derived. The alkanoic acid is renamed alkanamide. Any alkyl substituents on the nitrogen are named with the prefix *N*.
- Ethers and haloalkanes (alkyl halides) are named in a different way. The halogen atom in alkyl halides is named as a substituent (halo) of the main alkane chain. Ethers have two alkyl chains. The larger one is the main chain. The smaller one along with the oxygen atom is considered to be an alkoxy substituent of the main chain.
- Amines are named by placing the suffix -anamine after the root name. Any alkyl substituents on the nitrogen are named with the prefix *N*.

SECTION 6

Primary, Secondary, Tertiary and Quaternary Nomenclature

Primary, Secondary, Tertiary and Quaternary Nomenclature

The primary (1°), secondary (2°), tertiary (3°) and quaternary (4°) nomenclature is used in a variety of situations:

- to define a carbon centre.
- to define functional groups such as alcohols, halides, amines and amides.

6.1 Definition of carbon centres as 1°, 2°, 3° or 4°

One of the easiest ways of determining whether a carbon centre is 1°, 2°, 3° or 4° is to count the number of bonds which are *not* bonded to hydrogen.

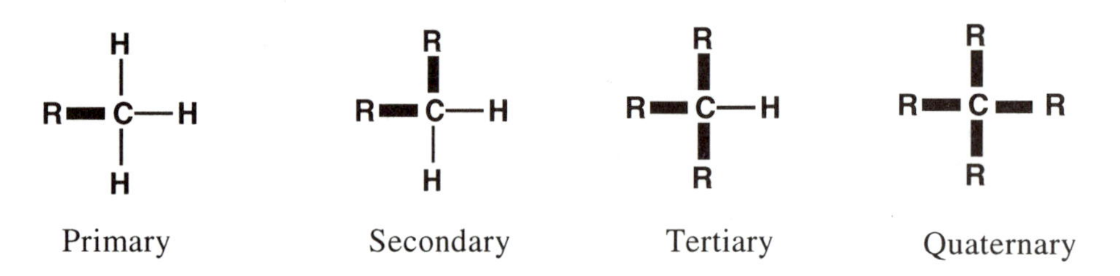

Identify whether the carbons in the following structure are primary, secondary, tertiary or quaternary.

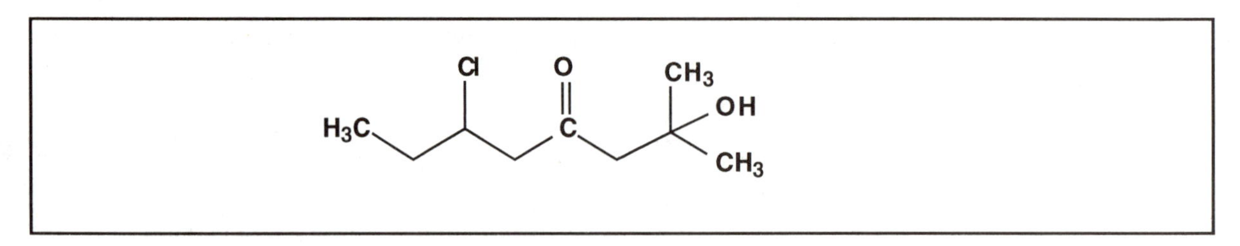

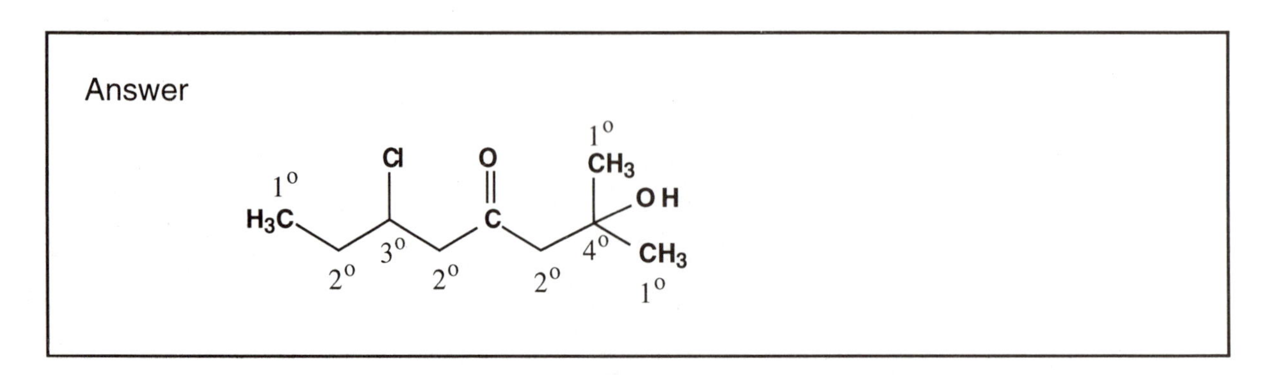

6.2 Nomenclature for amines and amides

Amines and amides can be defined as being primary, secondary, tertiary or quaternary depending on the number of bonds from **nitrogen** which do *not* lead to hydrogen.

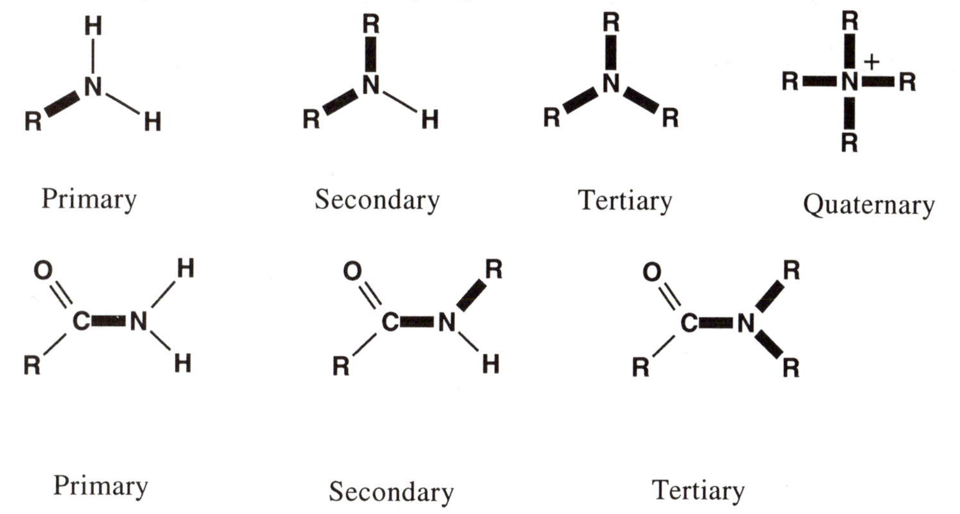

Identify the following amines and amides as primary, secondary, etc.

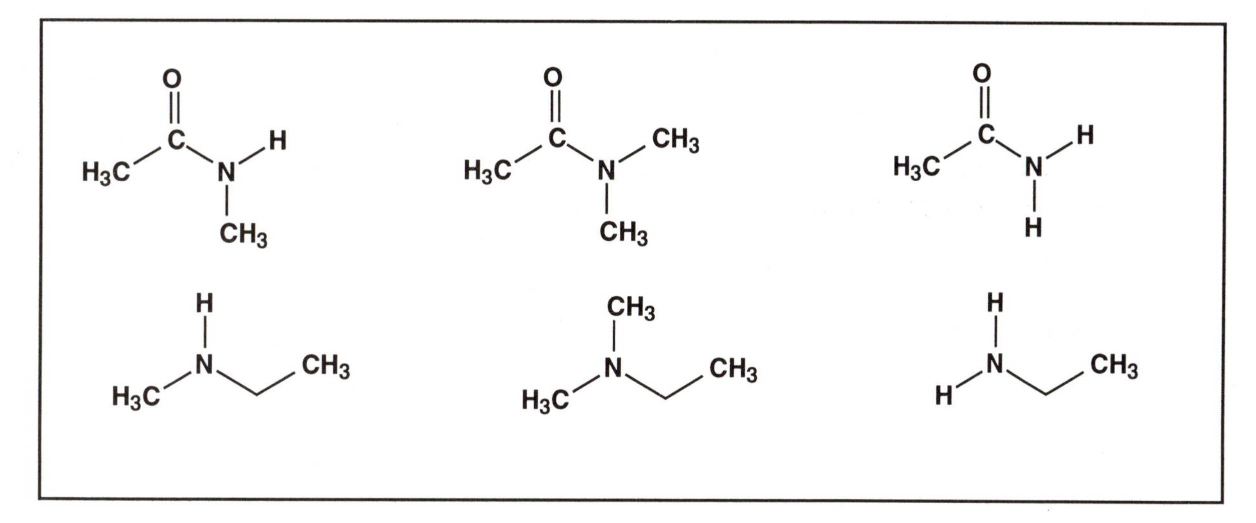

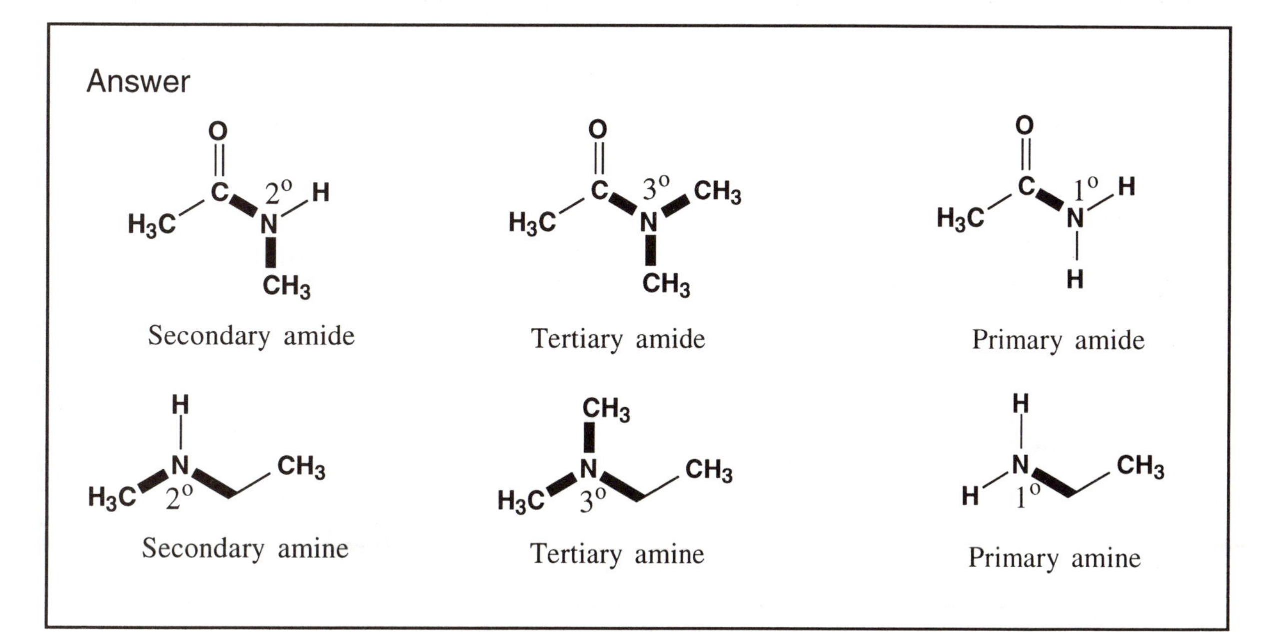

6.3 Nomenclature for alcohols and alkyl halides

Alcohols and alkyl halides are also defined as being primary, secondary or tertiary. However, the definition depends on the carbon to which the alcohol or halide is attached and it ignores the bond to the functional group. Thus, quaternary alcohols or alkyl halides are not possible.

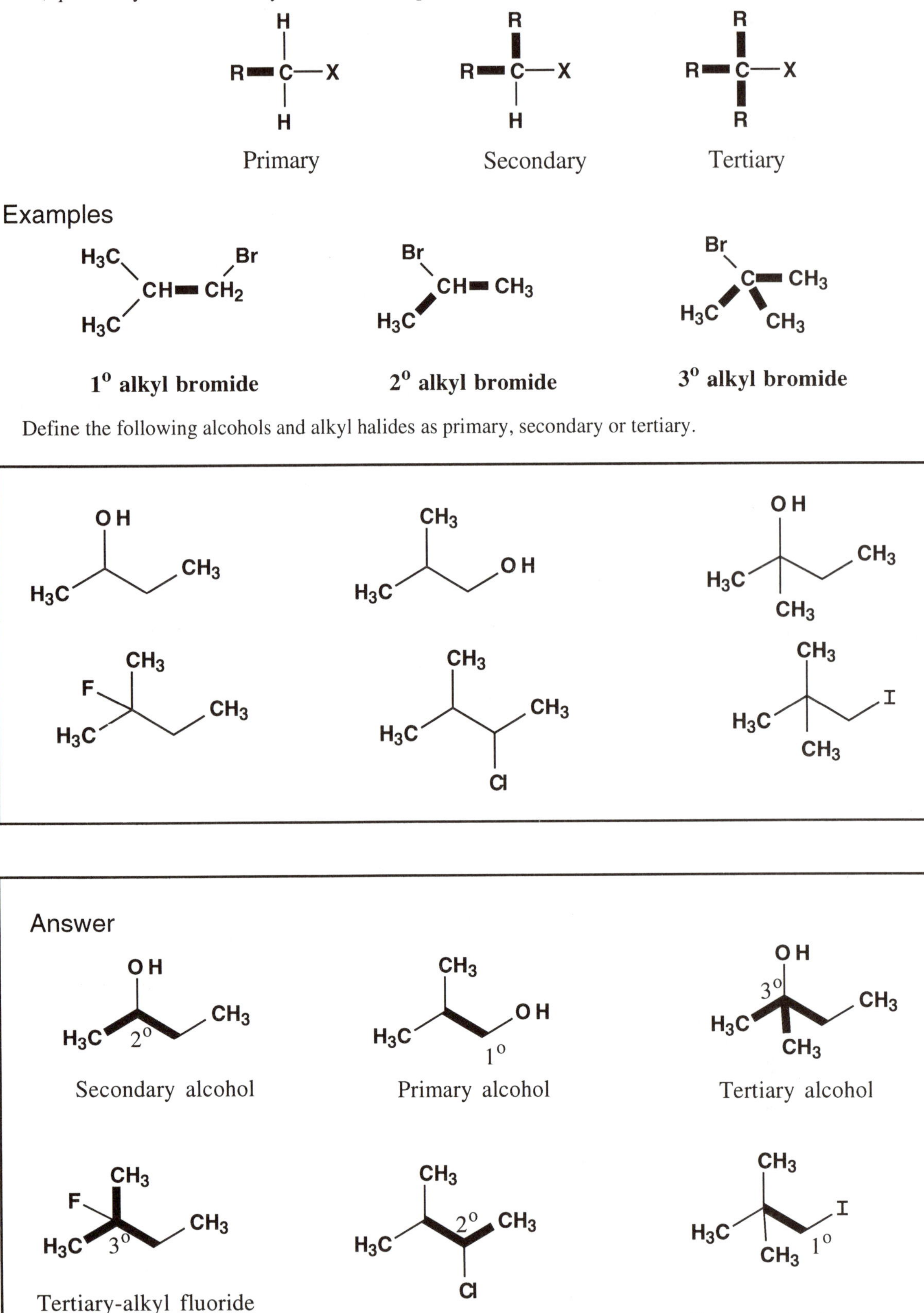

Note that the carbon to which a primary alcohol or primary alkyl halide is attached is actually a secondary carbon. Similarly, a secondary alcohol/alkyl halide is attached to a tertiary carbon and a tertiary alcohol/alkyl halide is attached to a quaternary carbon.

Summary

- Carbon centres are defined as primary, secondary, tertiary, or quaternary depending on the number of bonds not bonded to hydrogen.

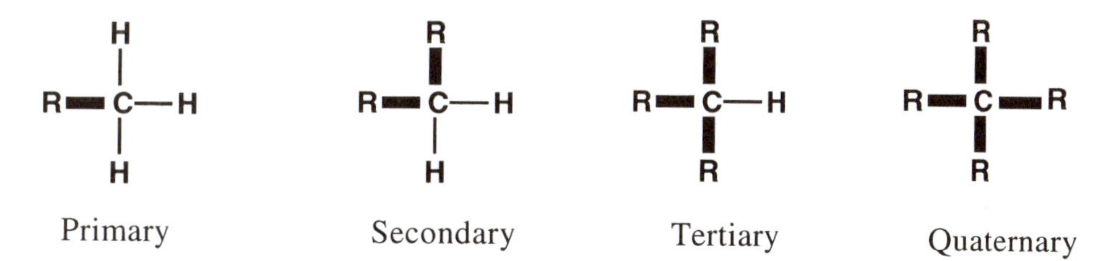

Primary Secondary Tertiary Quaternary

- Amines are defined as primary, secondary, tertiary or quaternary depending on the number of bonds from nitrogen not bonded to hydrogen.

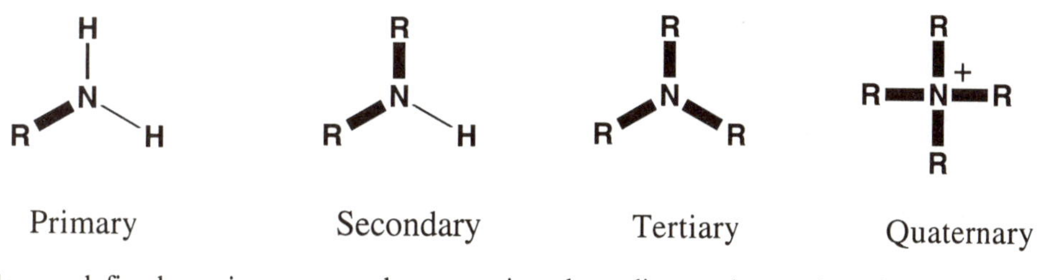

Primary Secondary Tertiary Quaternary

- Amides are defined as primary, secondary or tertiary depending on the number of bonds from nitrogen not bonded to hydrogen.

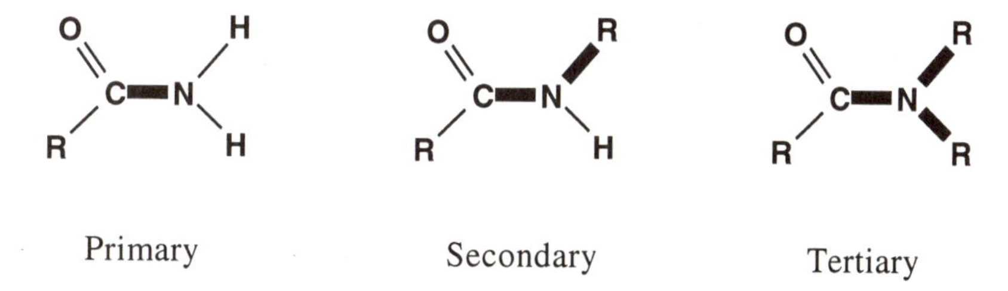

Primary Secondary Tertiary

- Quaternary amides are not possible.
- Alcohols and alkyl halides are defined as being primary, secondary, or tertiary depending on the nature of the bonds from the carbon bearing the OH group or halide atom. The number of bonds leading to groups other than hydrogen (and not including the bond to the functional group) determines the nomenclature.

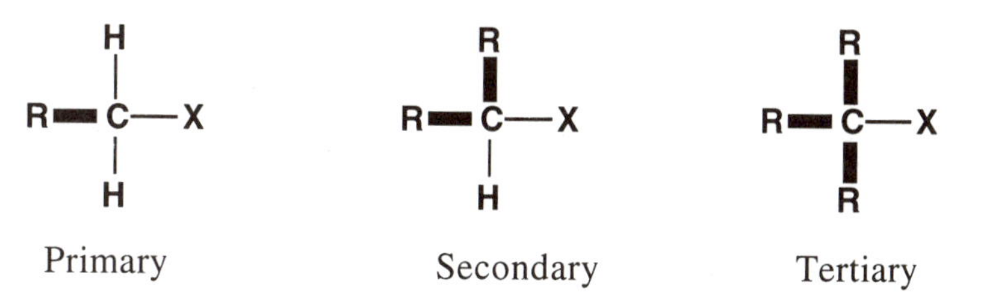

Primary Secondary Tertiary

- Quaternary alcohols and alkyl halides are not possible.

SECTION 7

Constitutional Isomers—Alkanes

Constitutional Isomers—Alkanes

7.1 Recognising constitutional isomers

Constitutional isomers are compounds which have the same molecular formula but have the atoms joined together in a different way. In other words, they have different carbon skeletons.

Examples

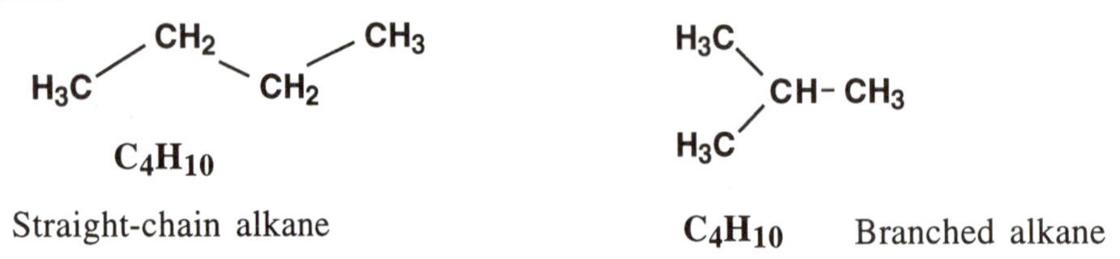

These are different compounds with different physical and chemical properties.

You may need practice at deciding whether two structures drawn on the page are isomers or the same compound.

Are the following structures identical or are they constitutional isomers?

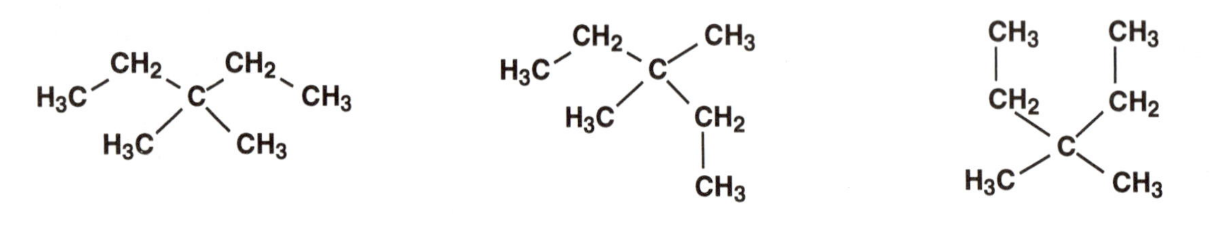

Answer

They are all identical. The same molecule has just been drawn in different ways. If you have trouble visualising this, then it would be a good idea to get into the habit of drawing structures such that the longest carbon chain is always written horizontally, as in the first diagram. Alternatively, buy a set of molecular models. If you are intending to continue with chemistry, these models are a worthwhile investment and are essential in later years.

Are the following molecules identical or are they constitutional isomers?

Answer

They are constitutional isomers. Both molecules have the same molecular formula (C_7H_{16}), but the atoms have been linked together differently.

7.2 Identifying consitutional isomers

Draw the structures of all the constitutional isomers having the molecular formula C_4H_{10}. (There are two)

Answer

This is a popular exam question. It is a good idea to tackle it in a systematic fashion so that you do not miss any isomers. First of all draw the simplest possible isomer. This is the isomer where the atoms are all linked in a single chain—a linear or unbranched alkane.

Now consider branched alkanes by shortening the chain by one carbon unit and placing a methyl group at different positions in the chain. In this example, there is only one possible branched isomer.

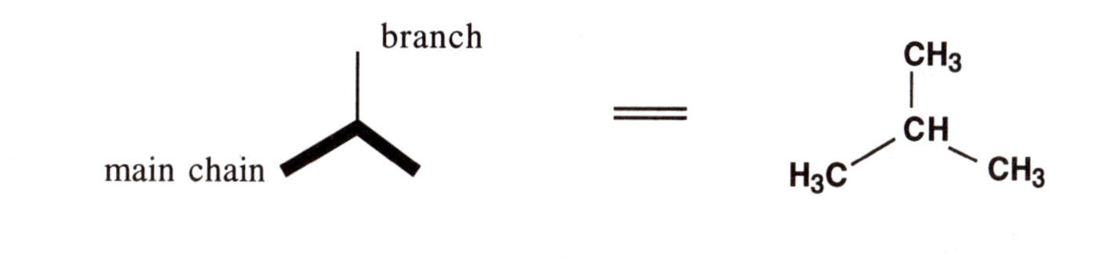

Draw the structures for all the constitutional isomers having the molecular formula C_5H_{12}.

Answer

The larger the molecule, the more isomers there are. Once again, tackle the problem systematically.
First of all, draw the straight-chain isomer.

Now shorten the chain by one unit and place the methyl group at different positions to see how many different branched isomers you get.

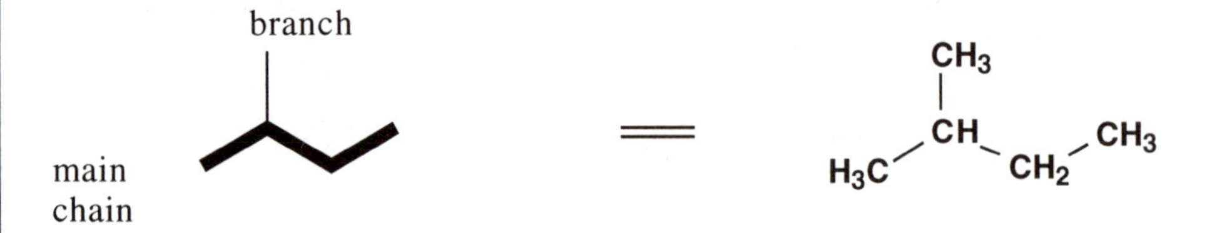

This is in fact the only possible branched isomer with a four-carbon main chain. If you put the branch at the other possible site, you end up with exactly the same molecule.

Reduce the main chain by another unit such that the main chain is three carbons long. You now have two methyl groups to put on. There is only one possible isomer.

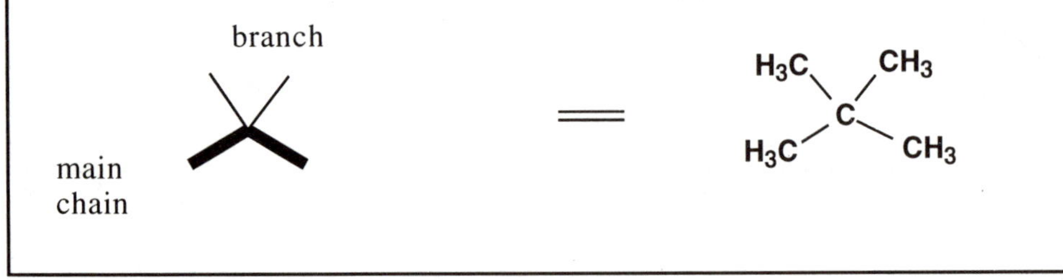

Summary

- Constitutional isomers are molecules having the same molecular formula but different carbon skeletons.
- Such isomers have different physical and chemical properties and are different compounds.
- The larger the molecule, the greater the number of isomers which are possible.
- To find all possible isomers for a particular molecular formula, start from the longest chain isomer, then cut the chain by one carbon unit and consider the number of branched isomers which are possible. Continue cutting the chain by one unit and working out the different branched isomers.

SECTION 8

Constitutional Isomers Containing Functional Groups

Constitutional Isomers Containing Functional Groups

8.1 Constitutional isomers containing a halide ($C_5H_{11}Cl$)

Earlier you did a problem identifying all the constitutional isomers of the formula C_5H_{12}. How many isomers would you get for the formula $C_5H_{11}Cl$?

Answer

This problem is a bit more complex than the earlier question on C_5H_{12}. With C_5H_{12} three isomers are possible. With $C_5H_{11}Cl$, eight isomers are possible. It is important to be systematic in answering this sort of question so that you don't miss any isomers. Firstly, identify the various carbon skeletons that are possible without considering where the chlorine goes. There are three possibilities, as follows:

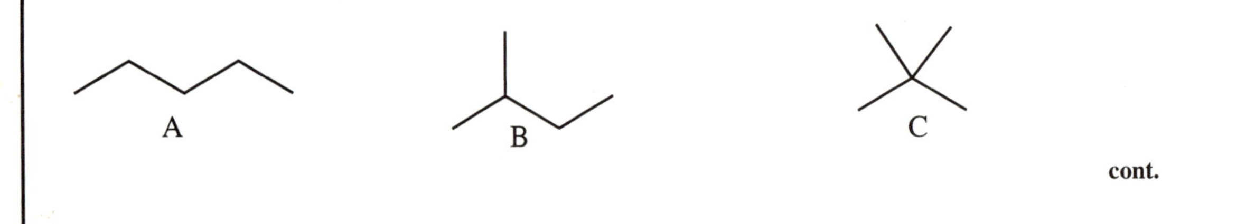

cont.

Now consider each of these structures in turn and consider the number of different places where the chlorine atom can be placed:

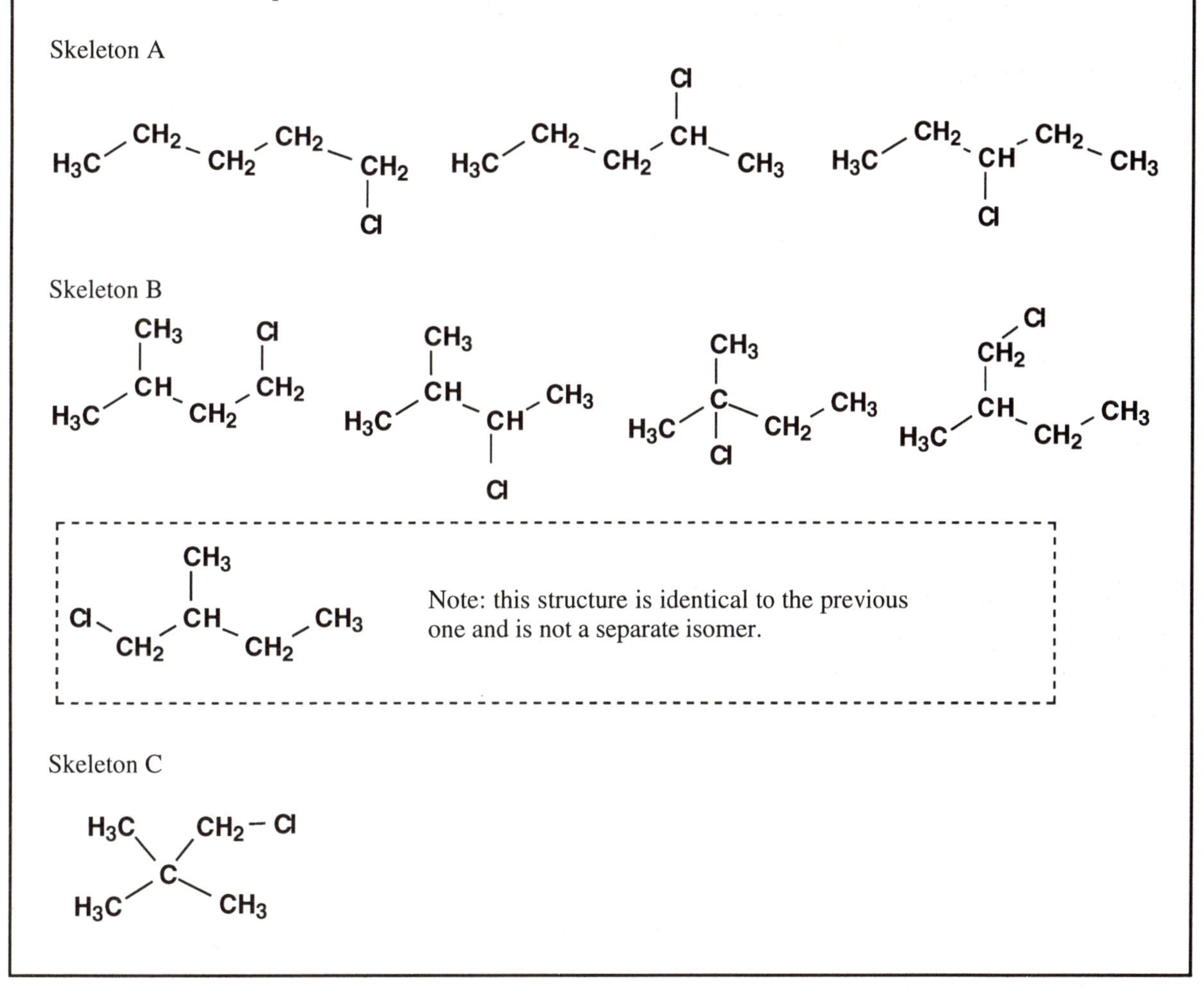

8.2 Constitutional isomers containing an alcohol ($C_4H_{10}O$)

Now draw the structures of all the possible alcohol isomers having the formula $C_4H_{10}O$.

Answer

The functional group is an alcohol (OH). There are two possible carbon skeletons:

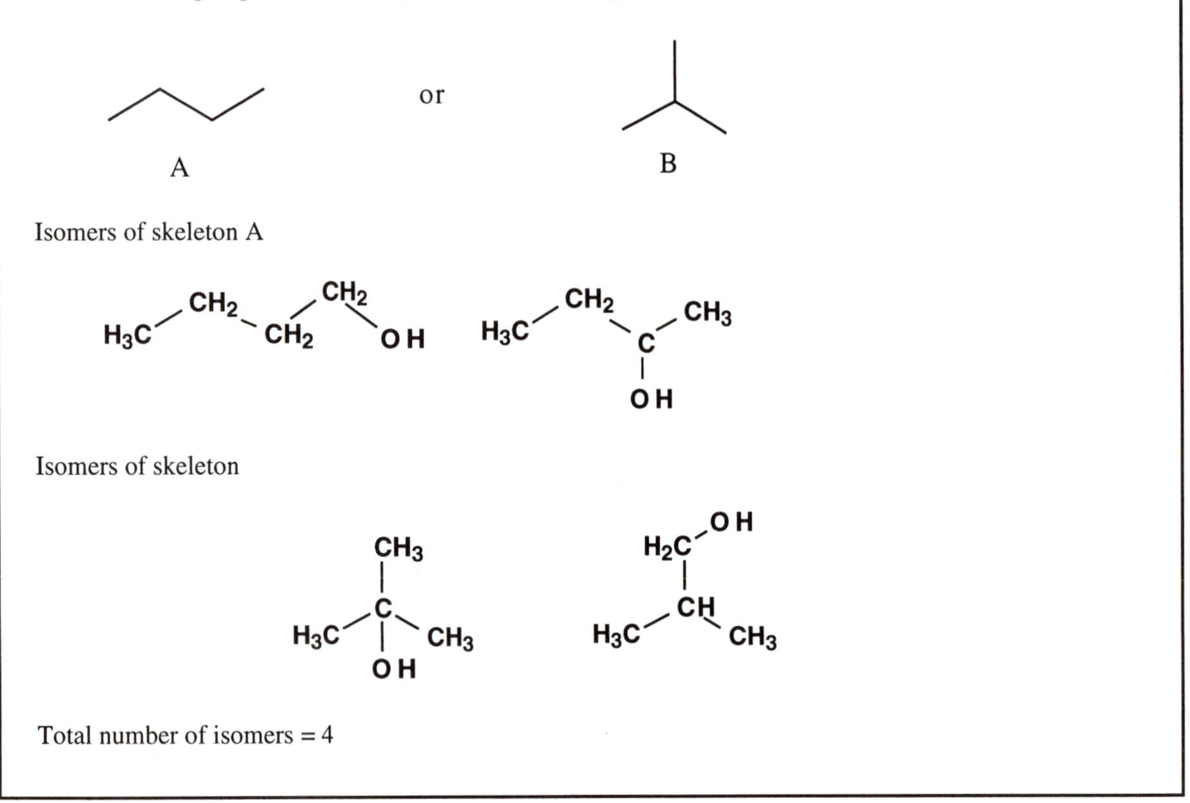

Total number of isomers = 4

8.3 Revision excercise on 1°, 2° and 3° nomenclature

Assign the alkyl halides and alcohols in the last two exercises as primary, secondary or tertiary.

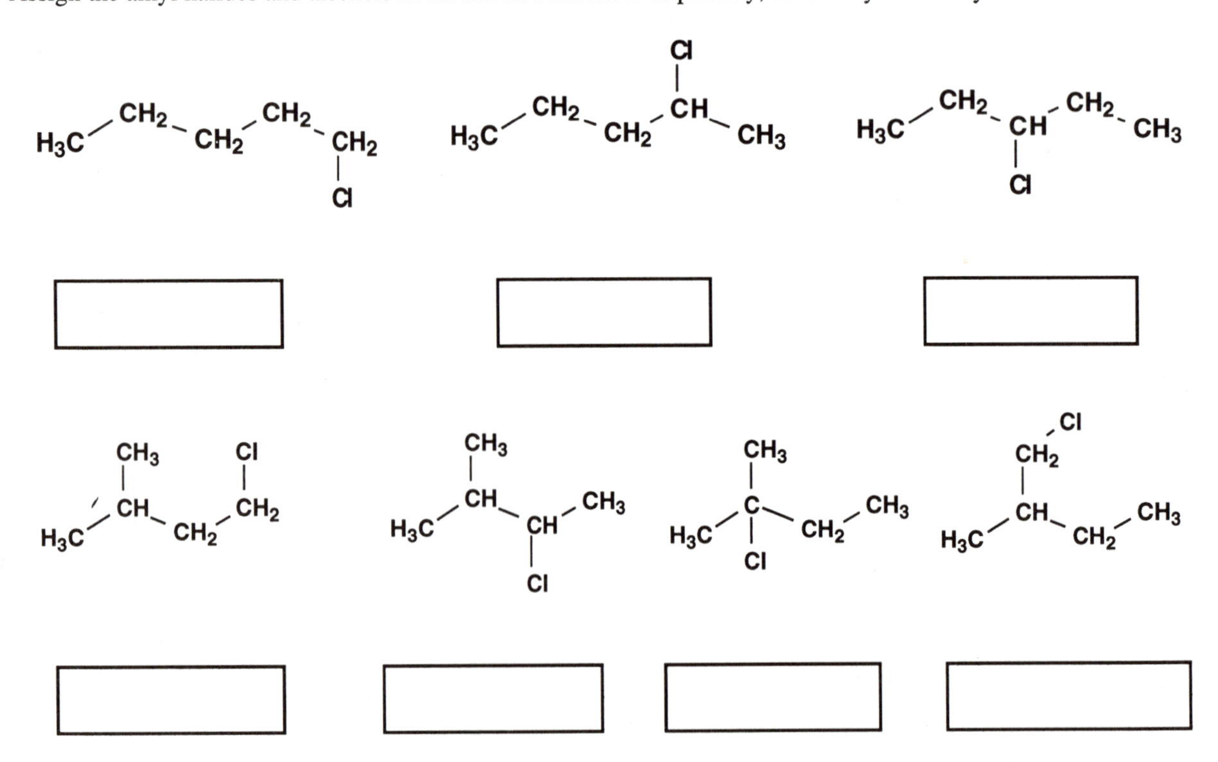

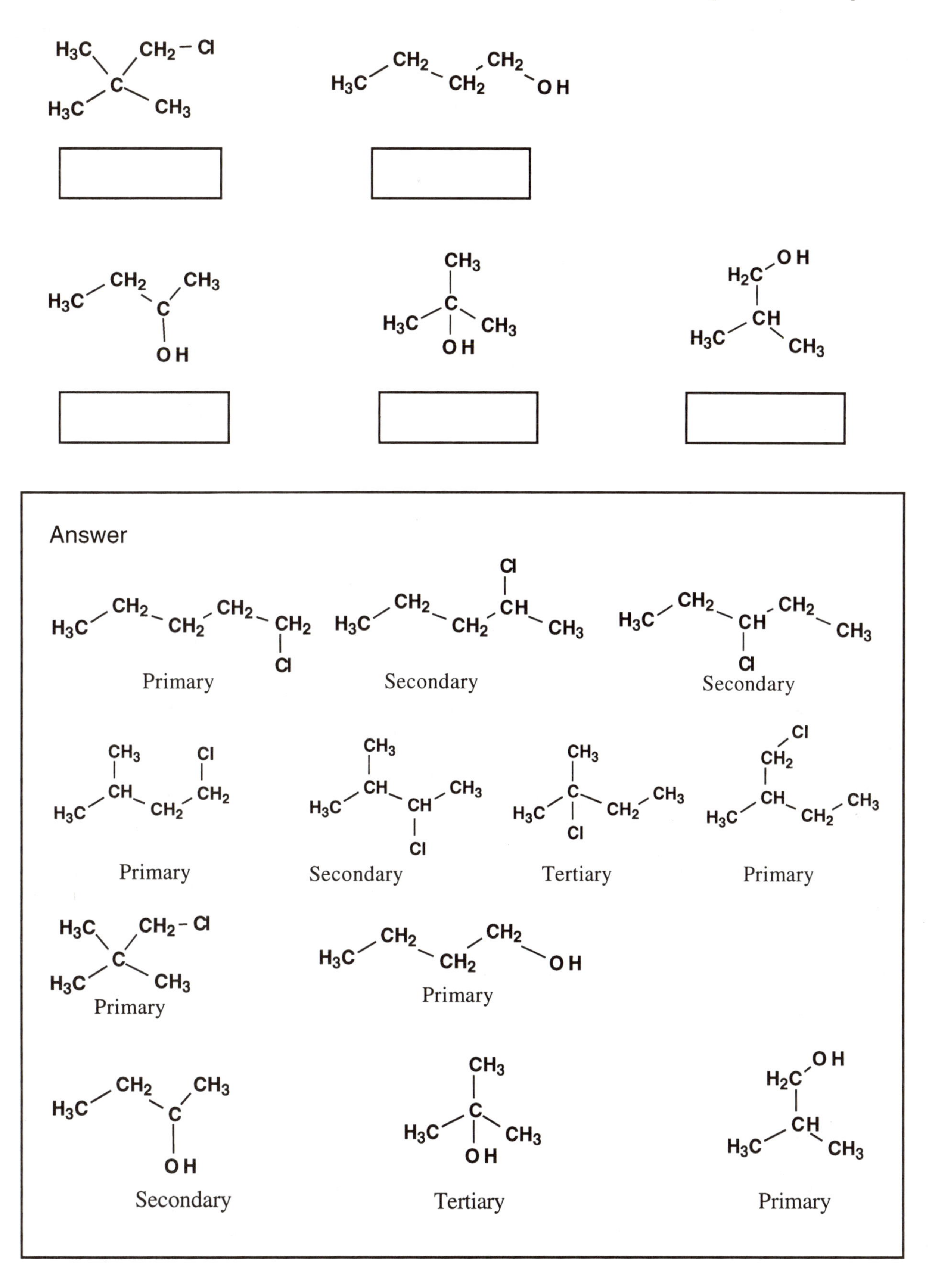

Summary

To work out the number of isomers of a structure containing a functional group, determine all the possible carbon skeletons first, then place the functional groups at different positions in each skeleton.

SECTION 9

Non-equivalent Carbons and Hydrogens 1

Non-equivalent Carbons and Hydrogens 1

9.1 Non-equivalent carbons and hydrogens

It is important to be able to identify equivalent and non-equivalent atoms in a chemical structure for a variety of reasons.

- Recognising equivalent and non-equivalent carbon atoms helps you to determine whether two molecules are identical or are isomers (see Sections 7 and 8).
- Recognising equivalent and non-equivalent hydrogens and carbons is crucial to the interpretation of the nuclear magnetic resonance spectrum (NMR) of a compound. (We are not covering NMR spectroscopy in this self-learning text, but you will come across it in your studies.)
- Non-equivalent carbons and hydrogens are likely to have differing reactivities.

In this section, we shall look at non-equivalent carbons and hydrogens for a variety of alkanes. In Section 10, we shall look at non-equivalent carbons and hydrogens in molecules containing a functional group.

9.2 Non-equivalent carbons in straight-chain alkanes

In order for carbons to be non-equivalent, there must be some definable difference between them. For example, in a straight-chain alkane, the carbon atoms within the chain are clearly different from those at the end of a chain. Similarly, carbons within the chain are non-equivalent depending on how far away they are from the end of the chain.

Identify the non-equivalent carbons in the following compounds.

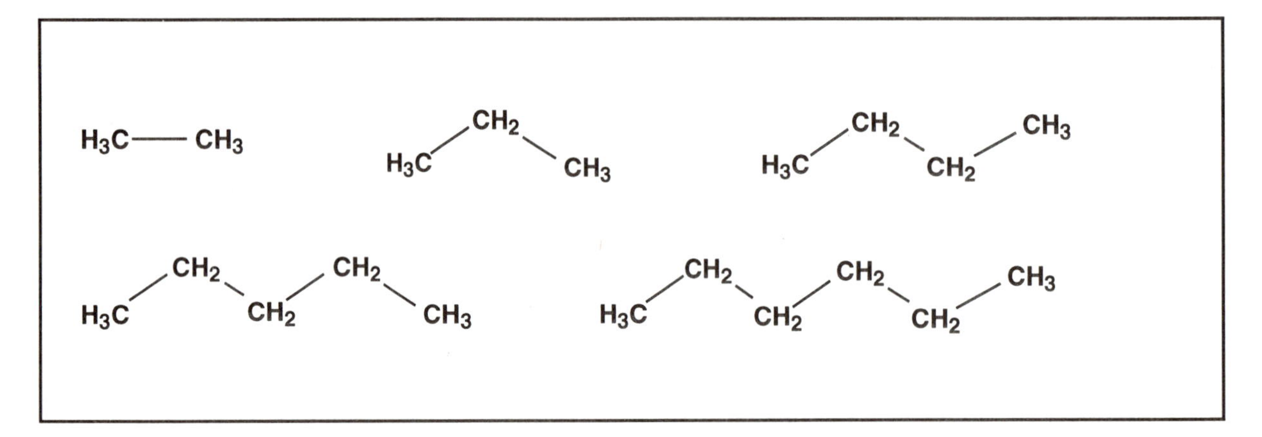

Answer

The non-equivalent carbons are labelled as shown.

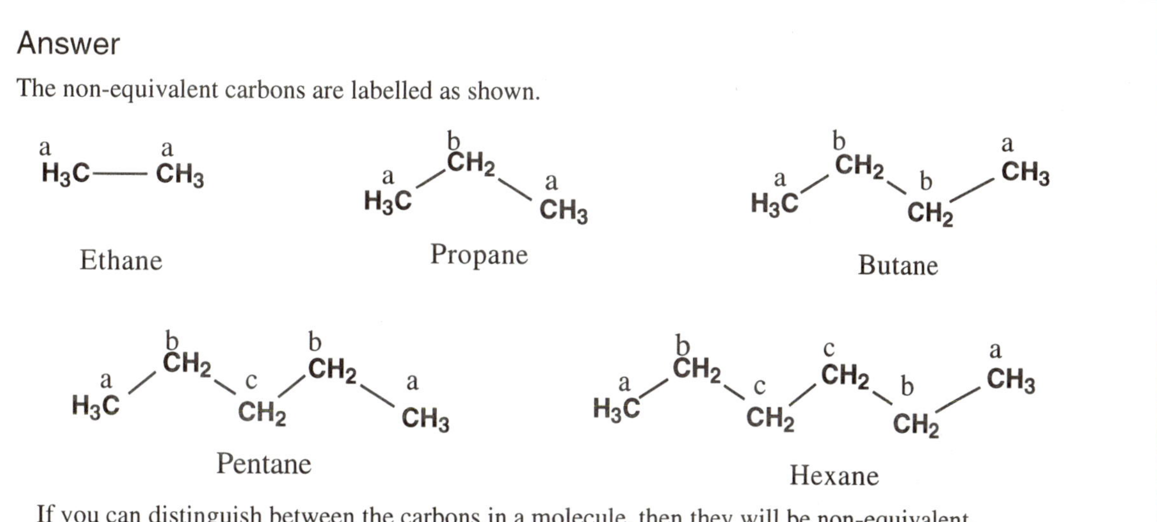

If you can distinguish between the carbons in a molecule, then they will be non-equivalent.

In ethane, the two carbons are equivalent. There is no way of distinguishing between them.

In propane, the middle carbon is distinct from the end carbons since it is a CH_2 unit compared to the end carbons which are CH_3 units.

In butane, the two inner carbons are equivalent CH_2 units and are distinct from the end carbons. The CH_3 end units are equivalent to each other.

In pentane, there are three CH_2 units, but the one in the middle of the molecule is different from the other two since it is further from the end of the chain.

Note that there is a difference between the terms 'similar' and 'equivalent'. The middle CH_2 group is similar to the CH_2 groups on either side. However, it is not equivalent to these groups.

9.3 Non-equivalent carbons in branched-chain alkanes

Introducing a branch into an alkane will often increase the number of sets of non-equivalent carbons.

Identify the non-equivalent carbons in the following branched alkanes.

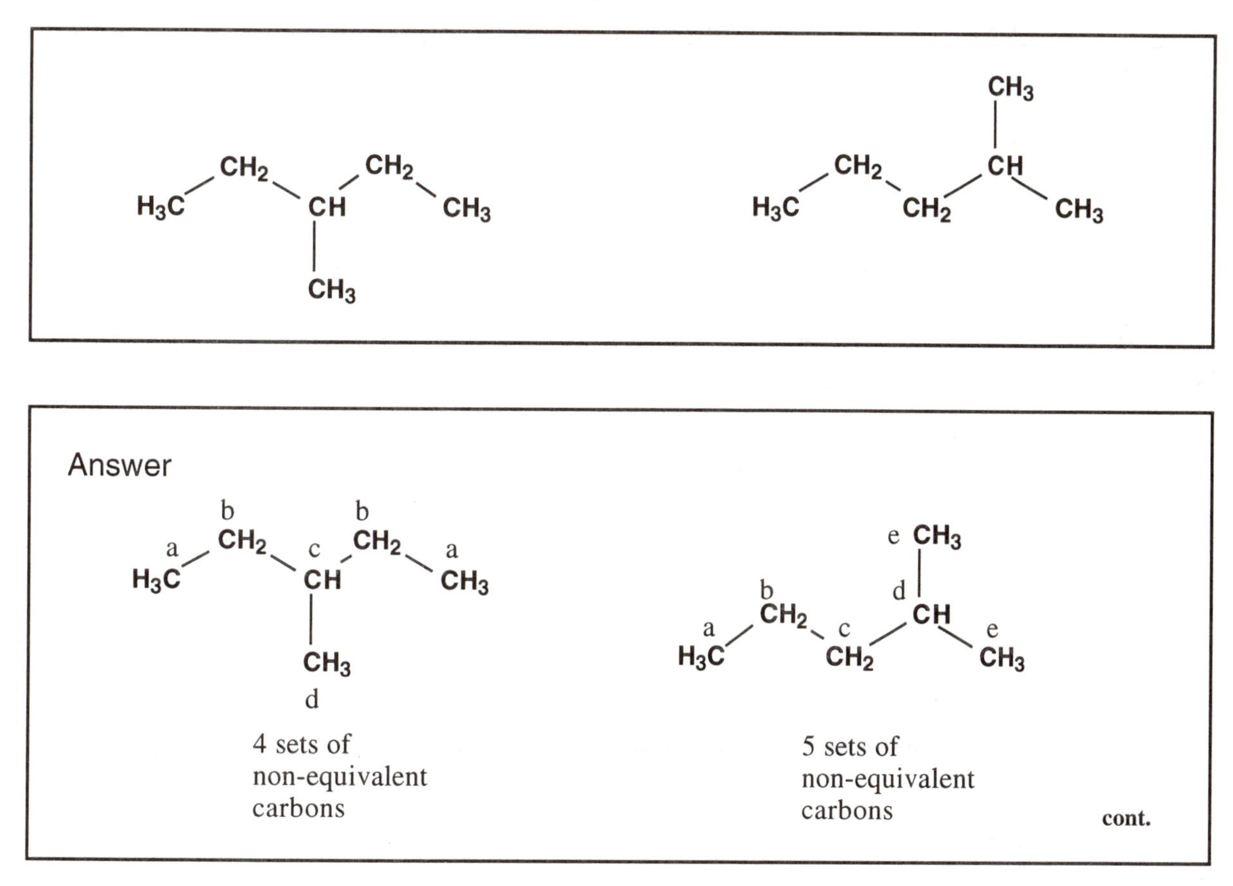

Note that the first structure has fewer sets of non-equivalent carbons because the the methyl branch is located in the centre of the molecule, introducing extra symmetry which is not present in the second structure.

Identify the non-equivalent carbons present in the following compounds.

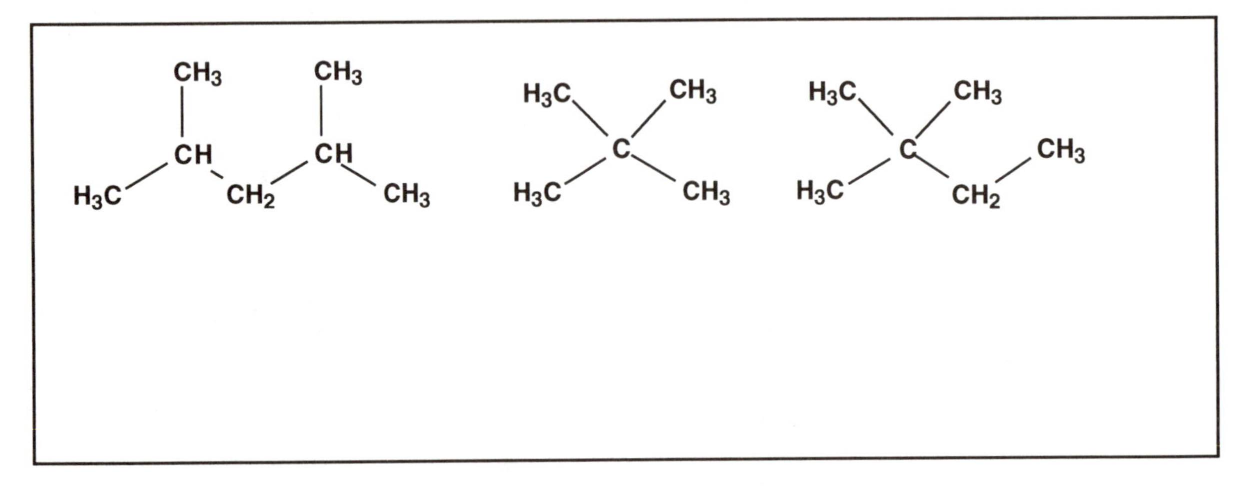

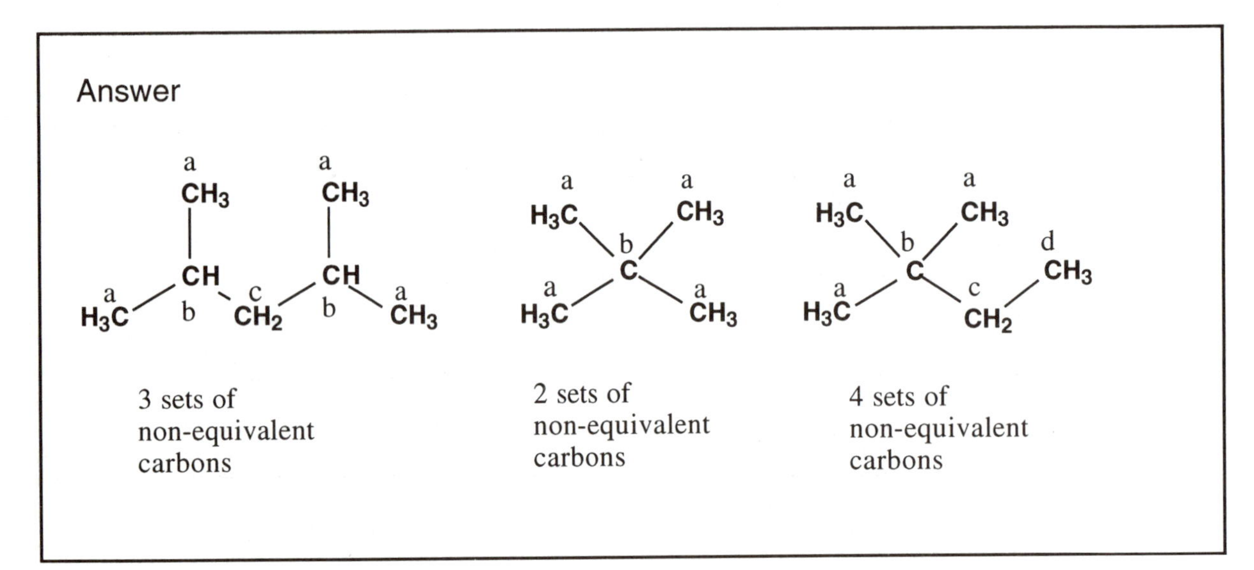

9.4 Non-equivalent hydrogens

There are more hydrogens than carbons to consider in an organic molecule, but a rule worth remembering for acyclic (non-cyclic) compounds is that the hydrogens on the same carbon will be equivalent. (There are exceptions to this rule, but we need not concern ourselves with these at this stage.) In other words, all the hydrogens on a CH_3 group will be equivalent. Similarly, both hydrogens on a CH_2 group will be equivalent.

A second rule worth noting for acyclic compounds is that the number of sets of non-equivalent hydrogens in a molecule will equal the number of sets of non-equivalent carbons *unless*:

(a) there is a carbon without any hydrogen attached to it, in which case there will be fewer sets of non-equivalent hydrogens; or

(b) a hydrogen is attached to another atom other than carbon (e.g. OH or NH_2), in which case there will be more sets of non-equivalent hydrogens.

Identify the non-equivalent hydrogens for the following straight-chain alkanes.

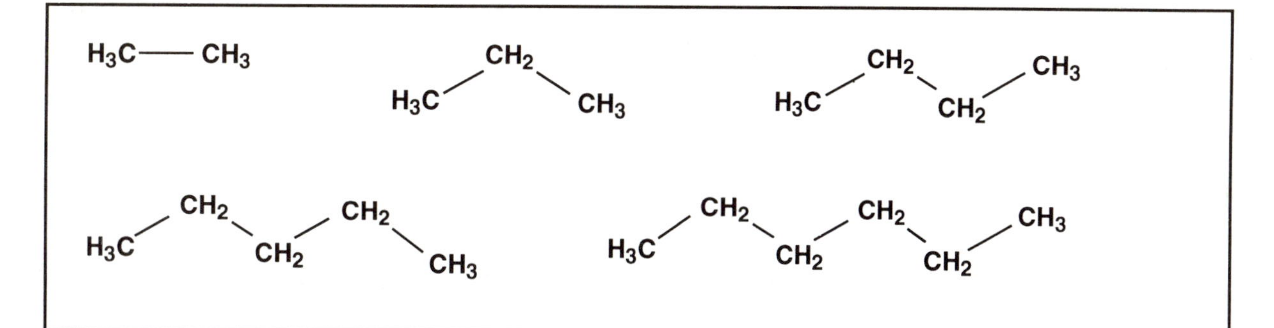

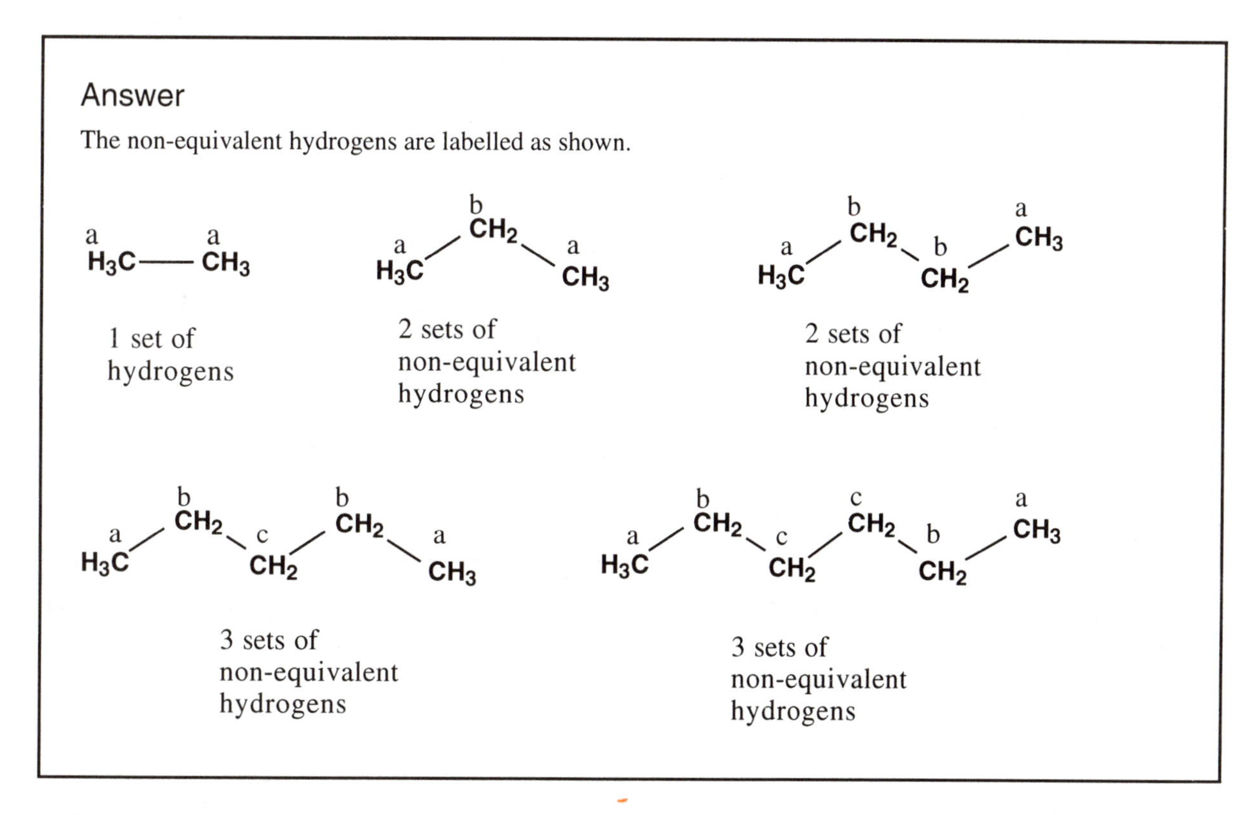

Identify the non-equivalent hydrogens in the following branched structures.

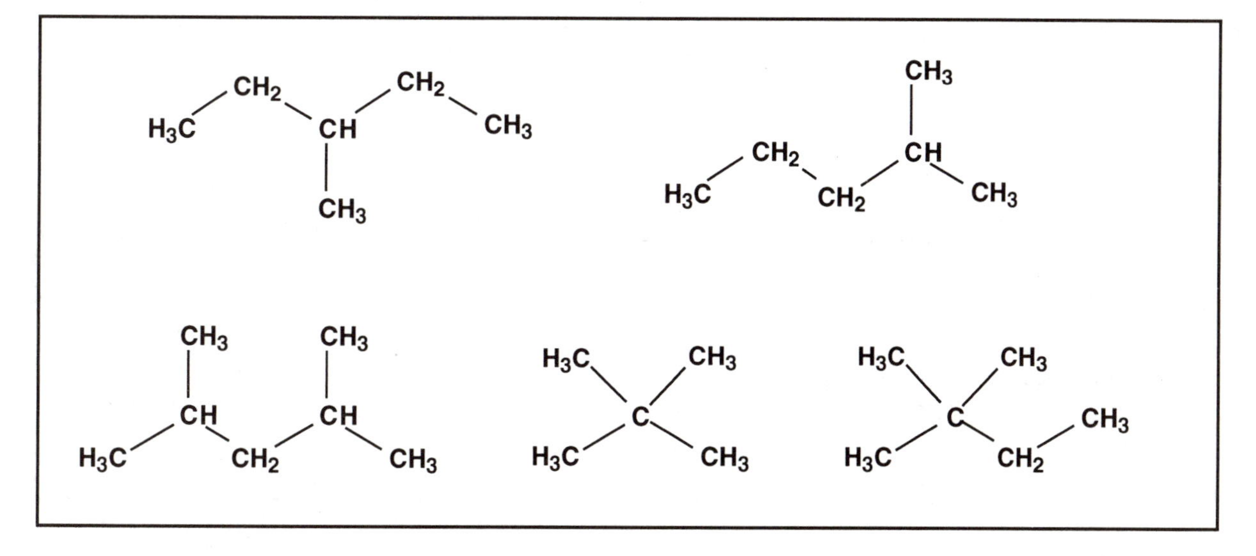

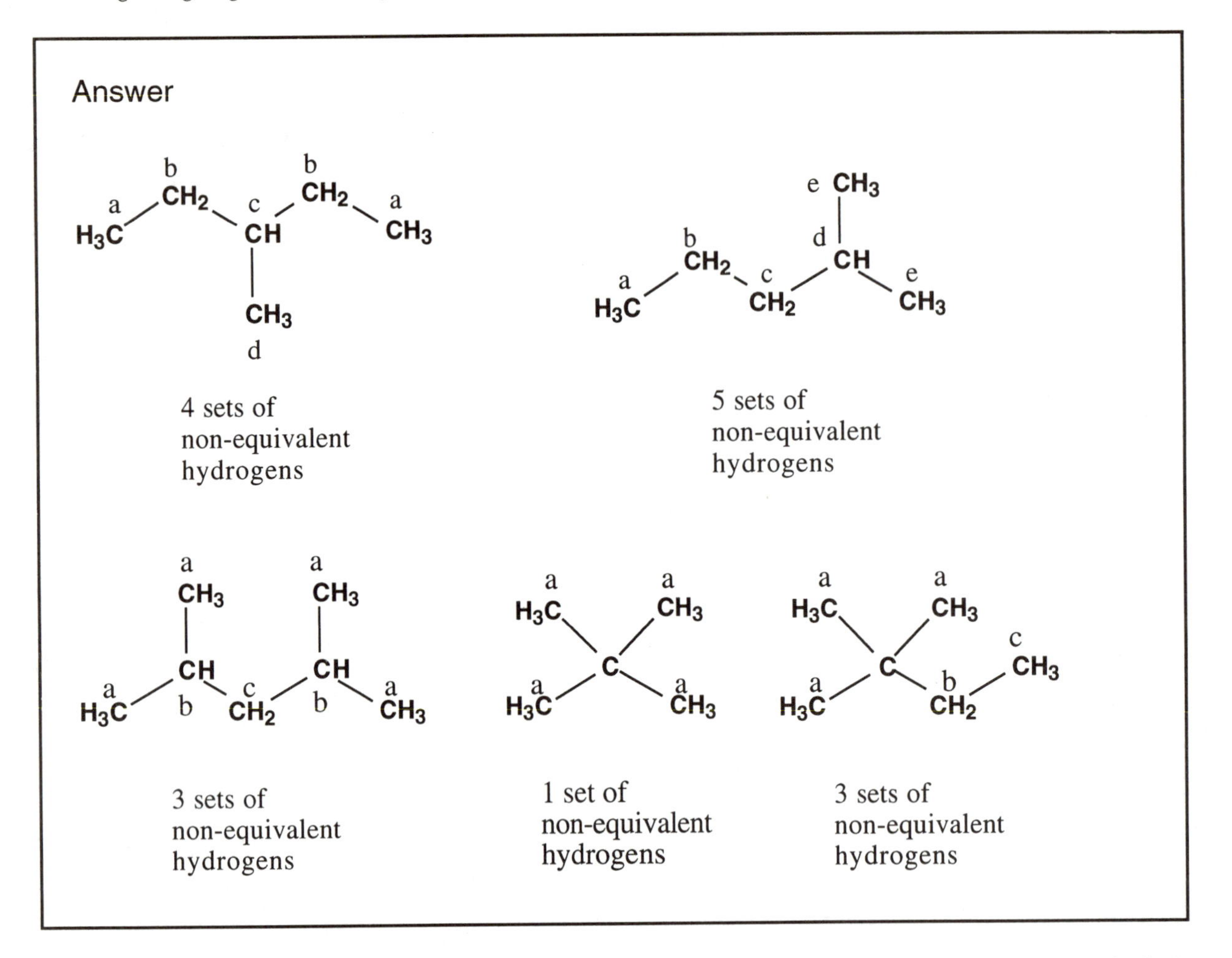

The structures above were the same ones used earlier in working out the number of sets of non- equivalent carbons. Compare the number of non-equivalent sets of carbons and non-equivalent hydrogens for each structure. Are they same? If not why not?

Answer

The number of sets of non-equivalent hydrogens and non-equivalent carbons is the same for all the structures except the last two. These structures have a carbon without a hydrogen attached and therefore the number of sets of non-equivalent hydrogens is one less than the number of sets of non-equivalent carbons.

Summary

- Equivalent carbons cannot be distinguished from each other.
- Equivalent hydrogens cannot be distinguished from each other.
- In general for acyclic compounds, hydrogens bonded to the same carbon are equivalent.
- The number of sets of non-equivalent hydrogens for a particular acyclic compound will be the same as the number of sets of non-equivalent carbons unless;

    (a) there is a carbon in the molecule without a hydrogen attached; or

    (b) there is a hydrogen bonded to an atom other than carbon.

- The number of sets of non-equivalent carbons or hydrogens will be greater for non symmetrical molecules than for symmetrical molecules.

SECTION 10

Non-equivalent Carbons and Hydrogens 2

Non-equivalent Carbons and Hydrogens 2

10.1 Non-equivalent atoms in hexane and 1-chlorohexane

In this section, we shall look at molecules containing a functional group.

Introducing a functional group into a straight-chain alkane can affect the number of non-equivalent carbons and hydrogens present by disrupting the symmetry of the molecule. Identify the non-equivalent hydrogens and carbons in the following structures.

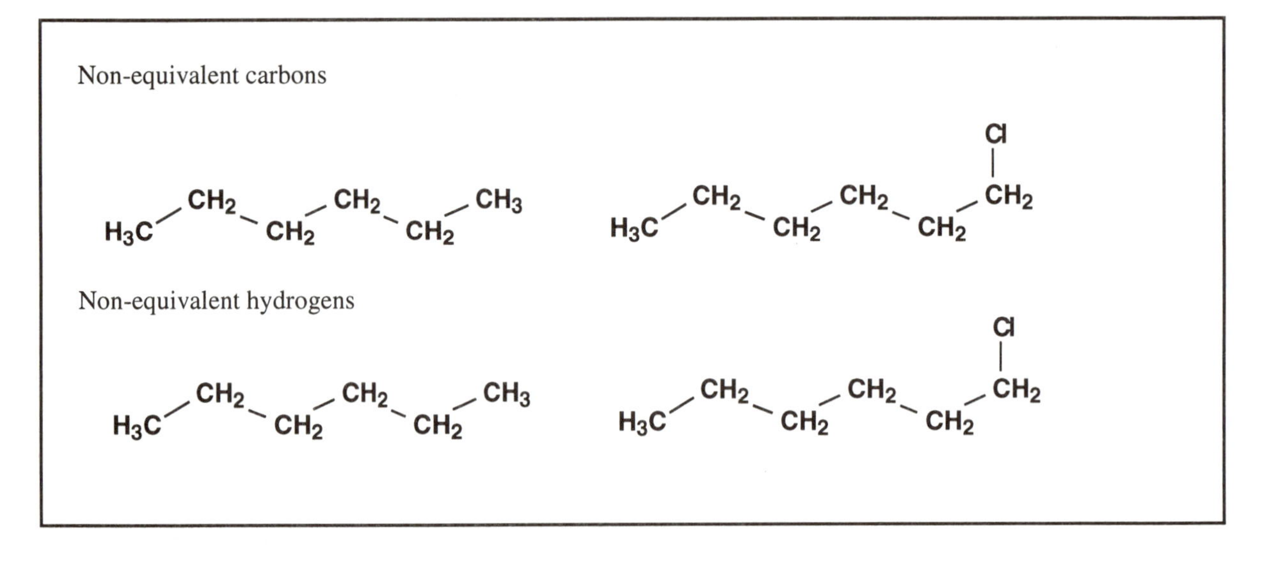

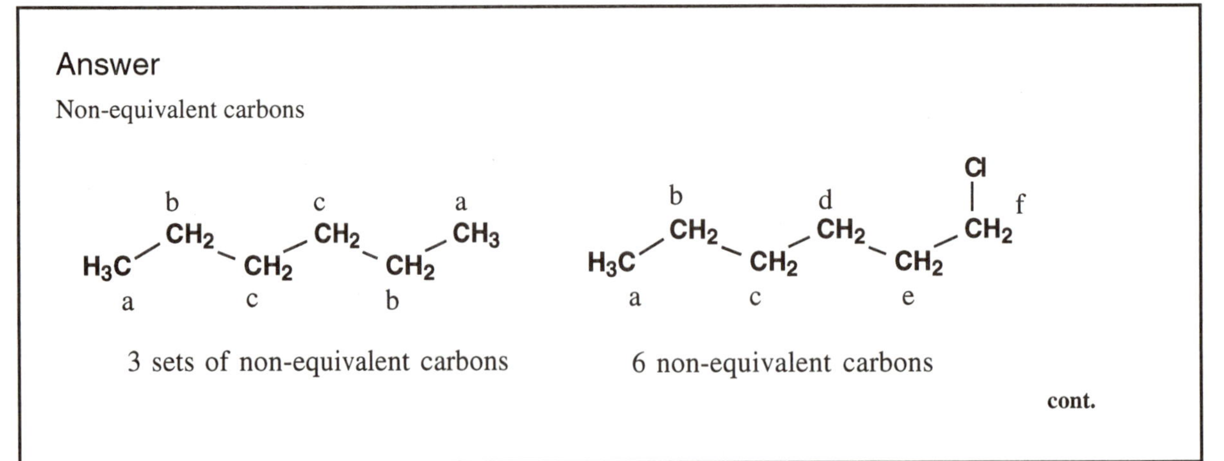

Non-equivalent hydrogens

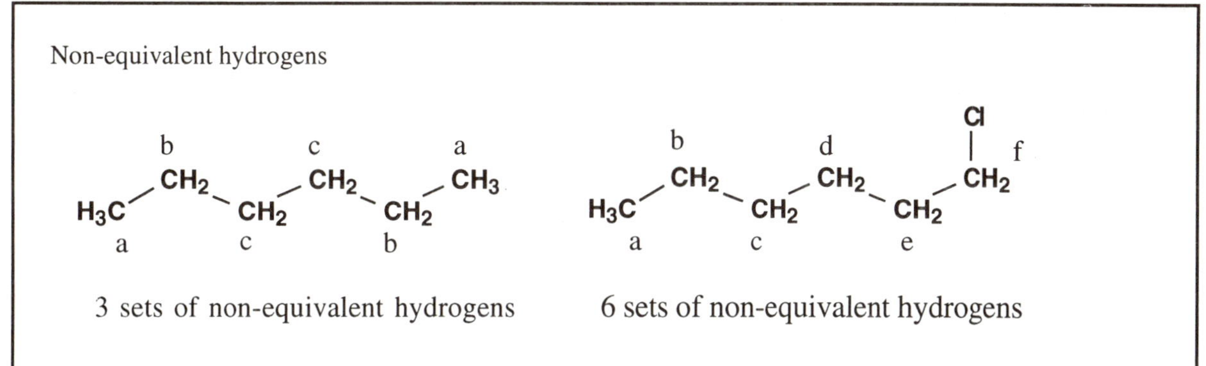

3 sets of non-equivalent hydrogens

6 sets of non-equivalent hydrogens

Note that the first molecule is symmetrical and therefore has only three different carbon atoms. The second molecule has a chloro group at one end which destroys the symmetry and so there are six non-equivalent carbon atoms. A common error is to see the two middle carbons as being equivalent, but carbon d is closer to the chlorine atom than carbon c and so the two carbons are different.

In this example, all the carbon atoms have hydrogens and all the hydrogens are on carbon atoms. Therefore, the number of non-equivalent hydrogens is the same as the number of non-equivalent carbons.

10.2 Non-equivalent atoms in alcohol isomers (C_5H_2O)

Draw the structures of the possible isomers of alcohols having the molecular formula $C_5H_{12}O$.

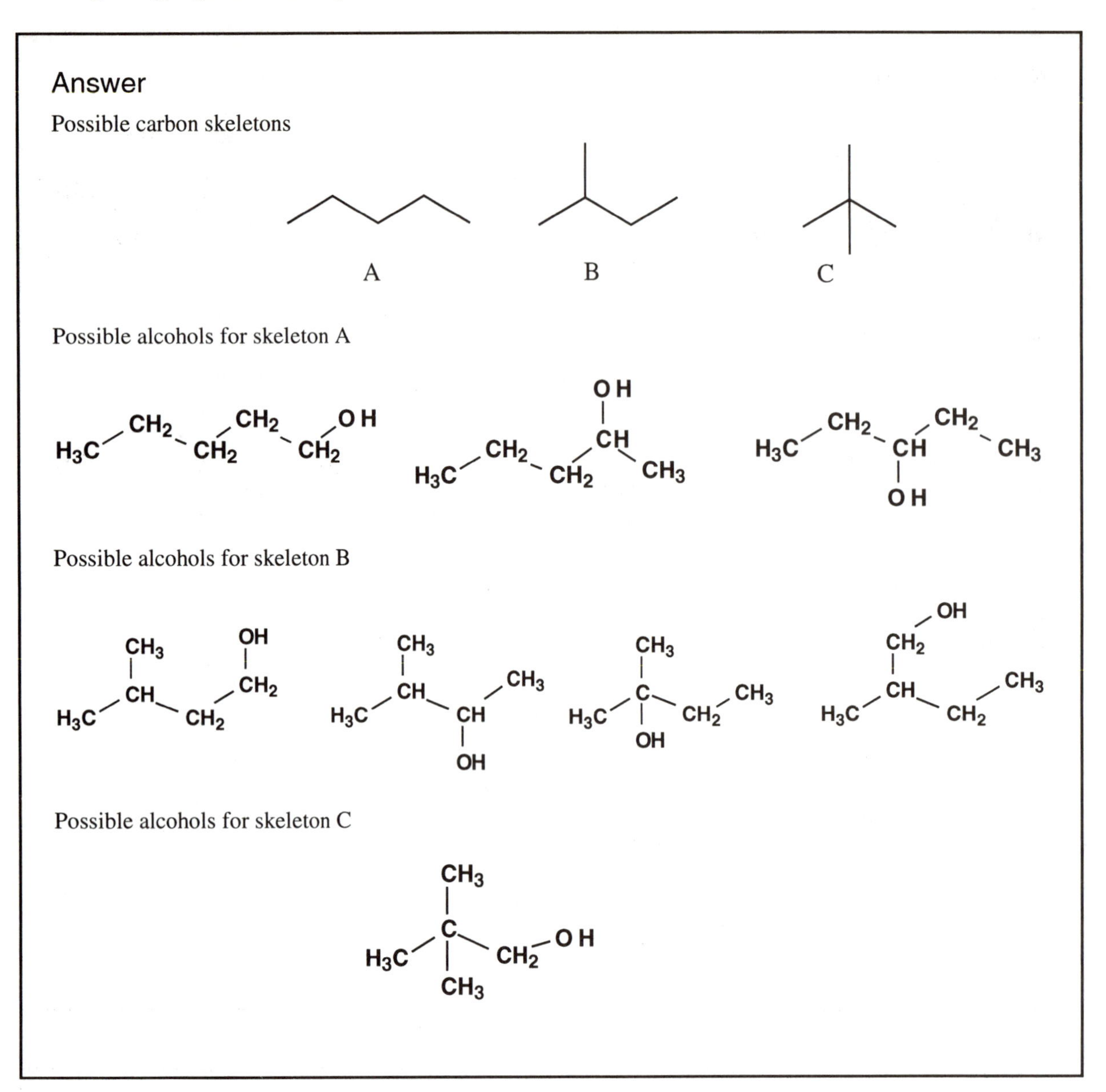

Identify the non-equivalent carbons and hydrogens for each of these isomers.

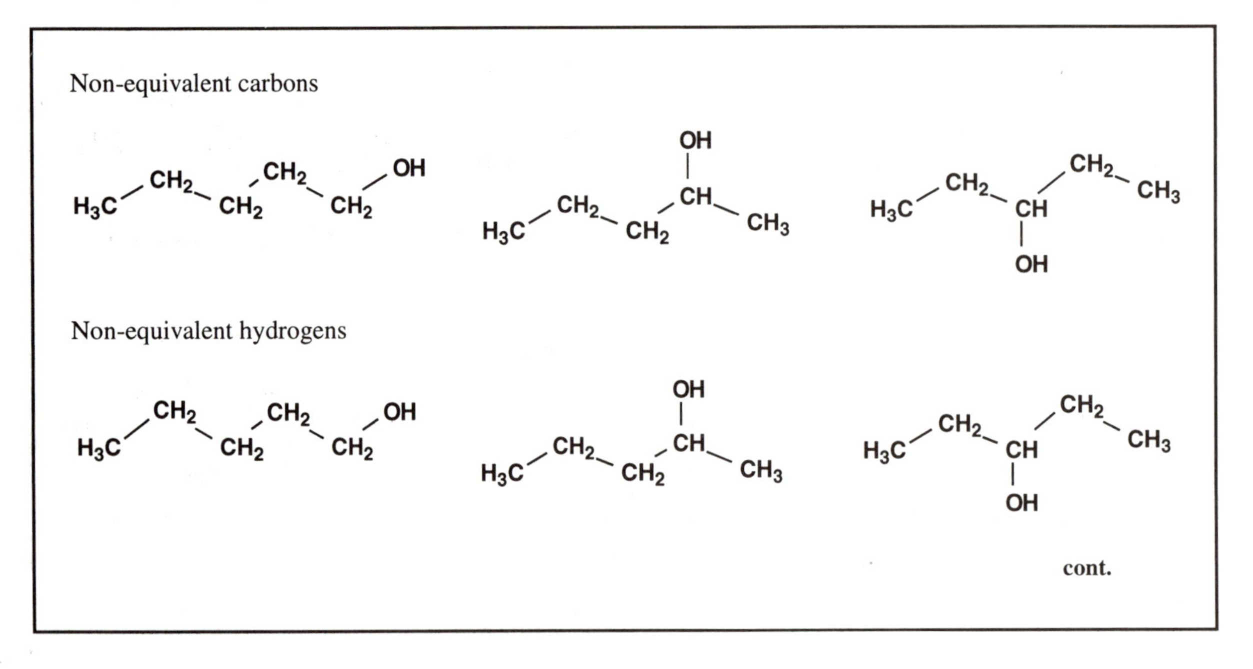

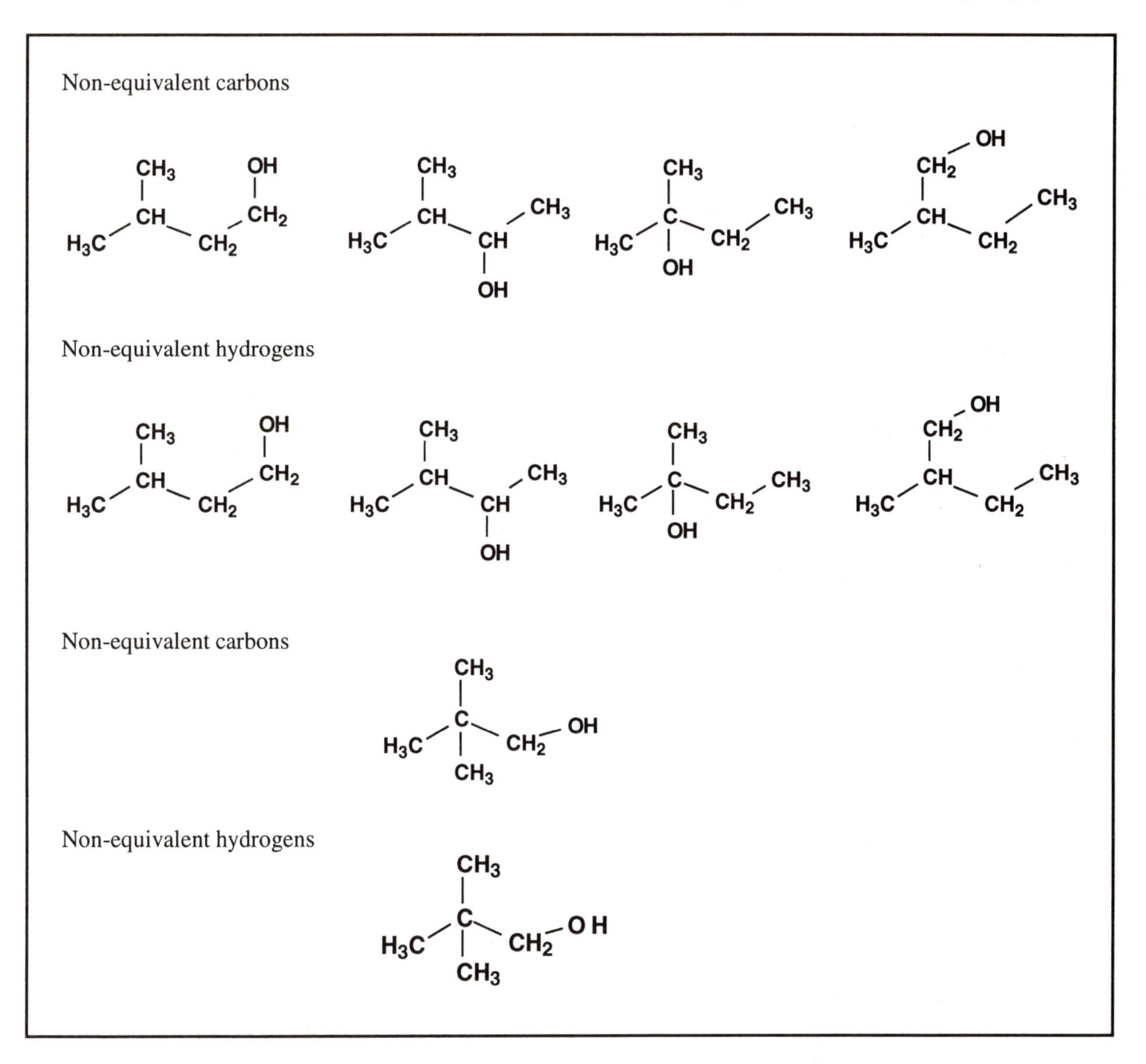
Non-equivalent carbons
Non-equivalent hydrogens
Non-equivalent carbons
Non-equivalent hydrogens

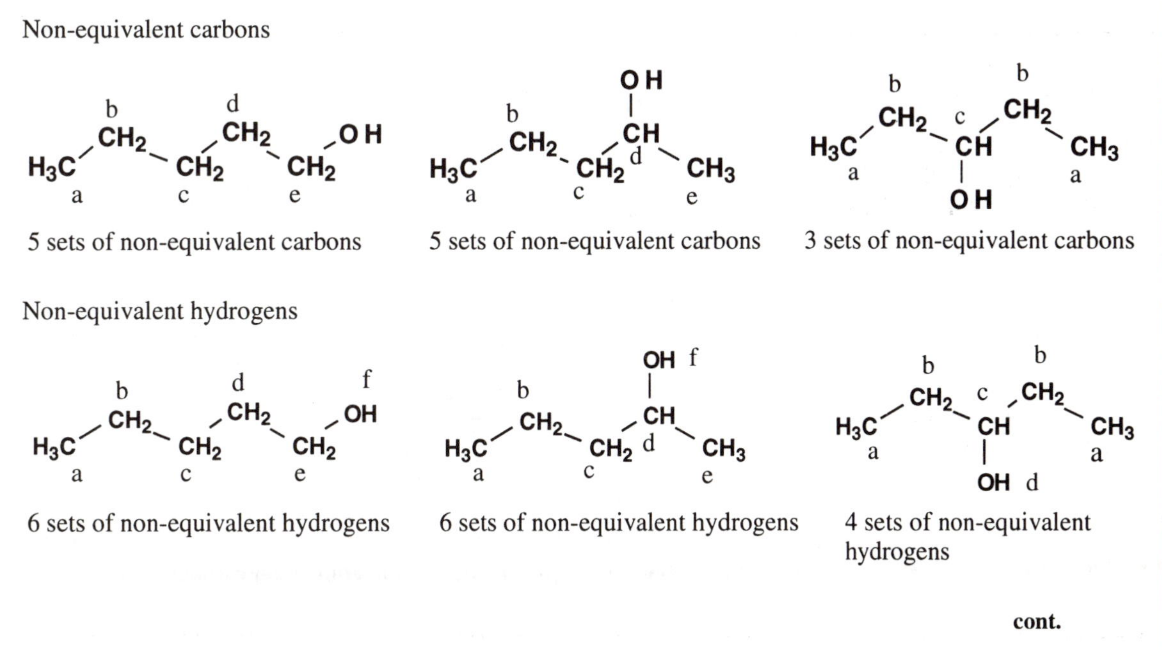
Answer
Non-equivalent carbons
5 sets of non-equivalent carbons
5 sets of non-equivalent carbons
3 sets of non-equivalent carbons
Non-equivalent hydrogens
6 sets of non-equivalent hydrogens
6 sets of non-equivalent hydrogens
4 sets of non-equivalent hydrogens
cont.

Notice that the third isomer above is symmetrical which means that there are only three non-equivalent carbon atoms present.

All the carbon atoms in the above structures have hydrogens attached and the non-equivalent hydrogens match up with the non-equivalent carbons. There is an extra hydrogen which is not attached to a carbon (i.e. OH) and so there is one extra non-equivalent hydrogen.

Non-equivalent carbons

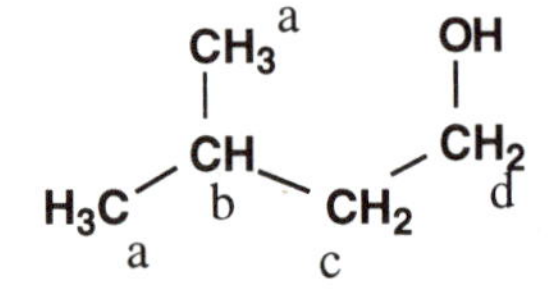

4 sets of non-equivalent carbons

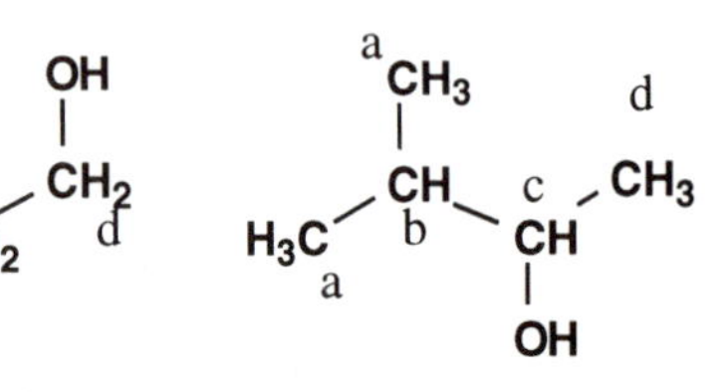

4 sets of non-equivalent carbons

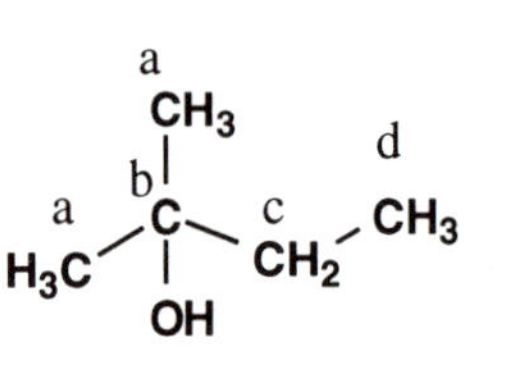

4 sets of non-equivalent carbons

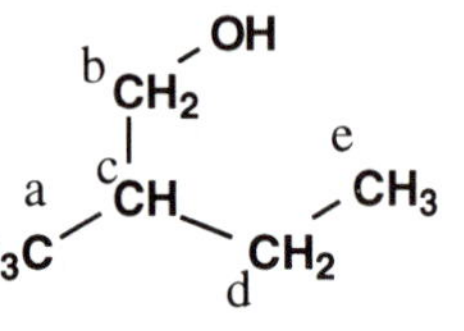

5 sets of non-equivalent carbons

Non-equivalent hydrogens

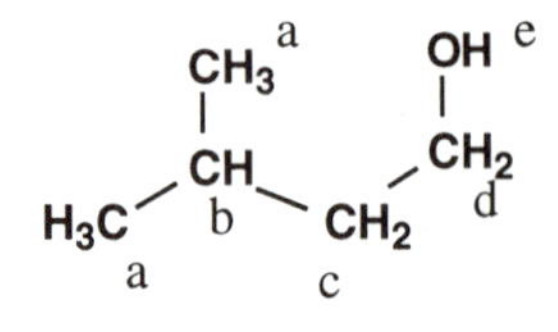

5 sets of non-equivalent hydrogens

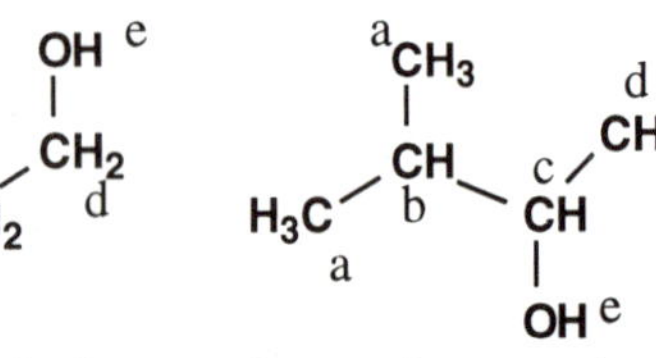

5 sets of non-equivalent hydrogens

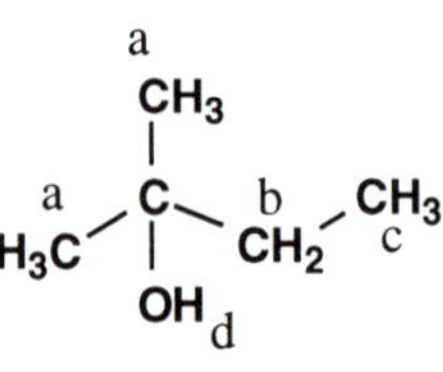

4 sets of non-equivalent hydrogens

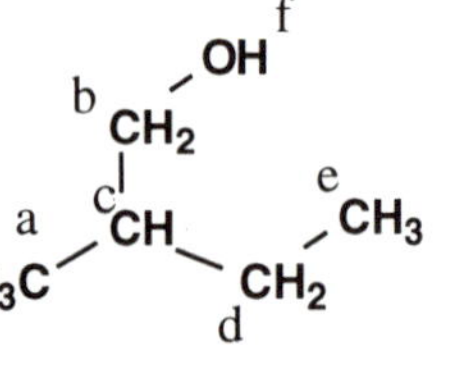

6 sets of non-equivalent hydrogens

Notice that the number of sets of non-equivalent hydrogens is one greater than the number of non-equivalent carbons due to the OH group. The exception is the third isomer above and that is because there is a quaternary carbon (i.e. a carbon without a hydrogen).

Non-equivalent carbons

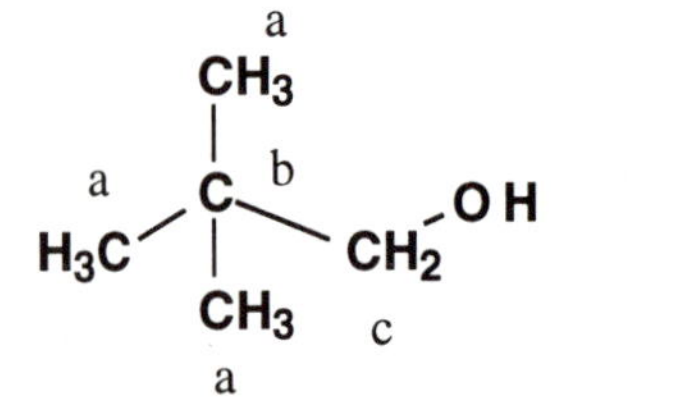

3 sets of non-equivalent carbons

Non-equivalent hydrogens

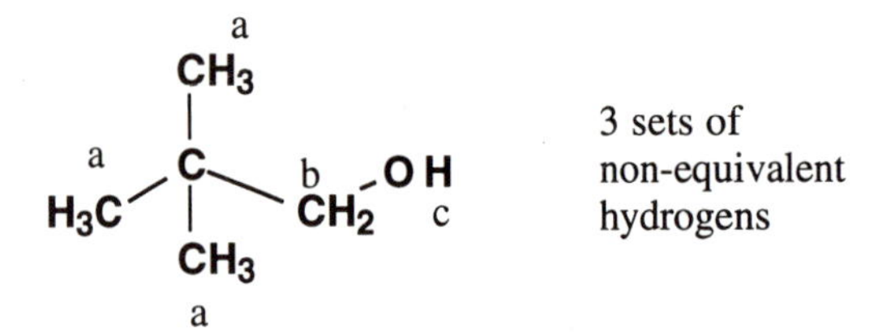

3 sets of non-equivalent hydrogens

There is one quaternary carbon (i.e. a carbon without a hydrogen) in this structure as well as the OH group, so that the number of non-equivalent hydrogens equals the number of non-equivalent carbons.

10.3 Non-equivalent atoms in cyclohexanone and 2-chlorocylohexanone

Identify the number of non-equivalent hydrogens and non-equivalent carbons in the following structures.

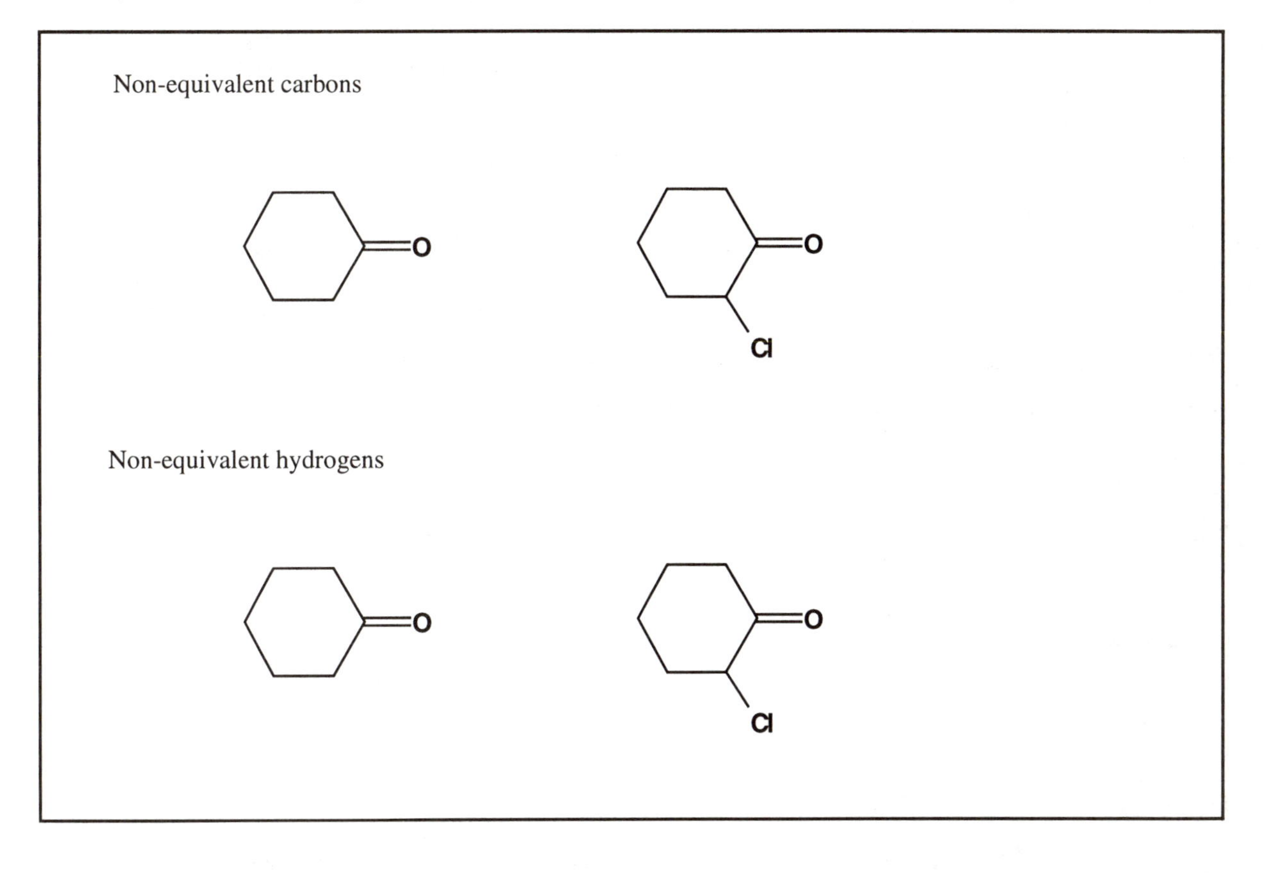

Answer

Non-equivalent carbons

The first structure is a symmetrical structure and will therefore have fewer non-equivalent carbons than the second structure which is not symmetrical. All the carbons in the second structure are different. If this is not obvious at first sight, count how many bonds each carbon is from the chlorine atom and then repeat the exercise to find out how many bonds each carbon is away from the carbonyl group. No carbon atom is the same. You may find two carbons which are the same number of bonds from the chlorine atom, but they won't be the same number of bonds away from the carbonyl bond. Hence, they are non-equivalent.

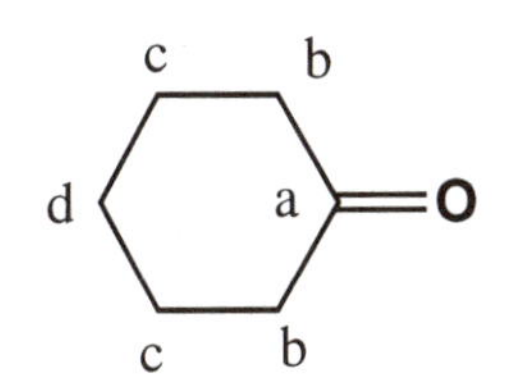

4 sets of non-equivalent carbons

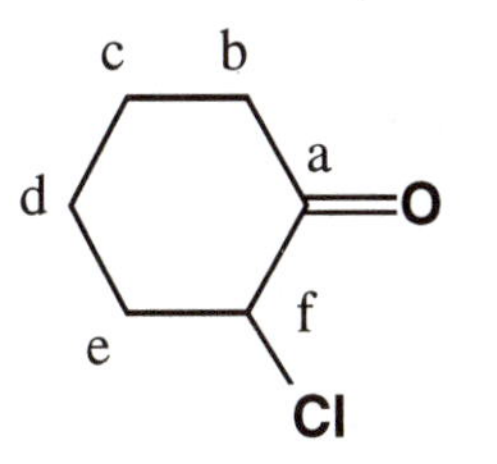

6 sets of non-equivalent carbons

Non-equivalent hydrogens

The number of sets of non-equivalent hydrogens is one less since the carbonyl carbon does not have any hydrogens attached to it.

cont.

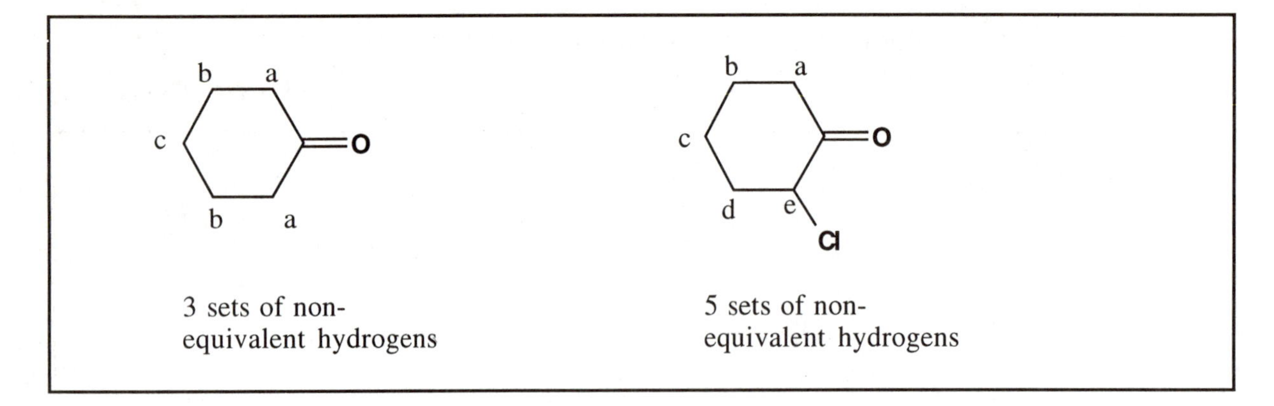

10.4 Non-equivalent atoms in miscellaneous structures

Identify the non-equivalent carbons and non-equivalent hydrogens in the following structures.

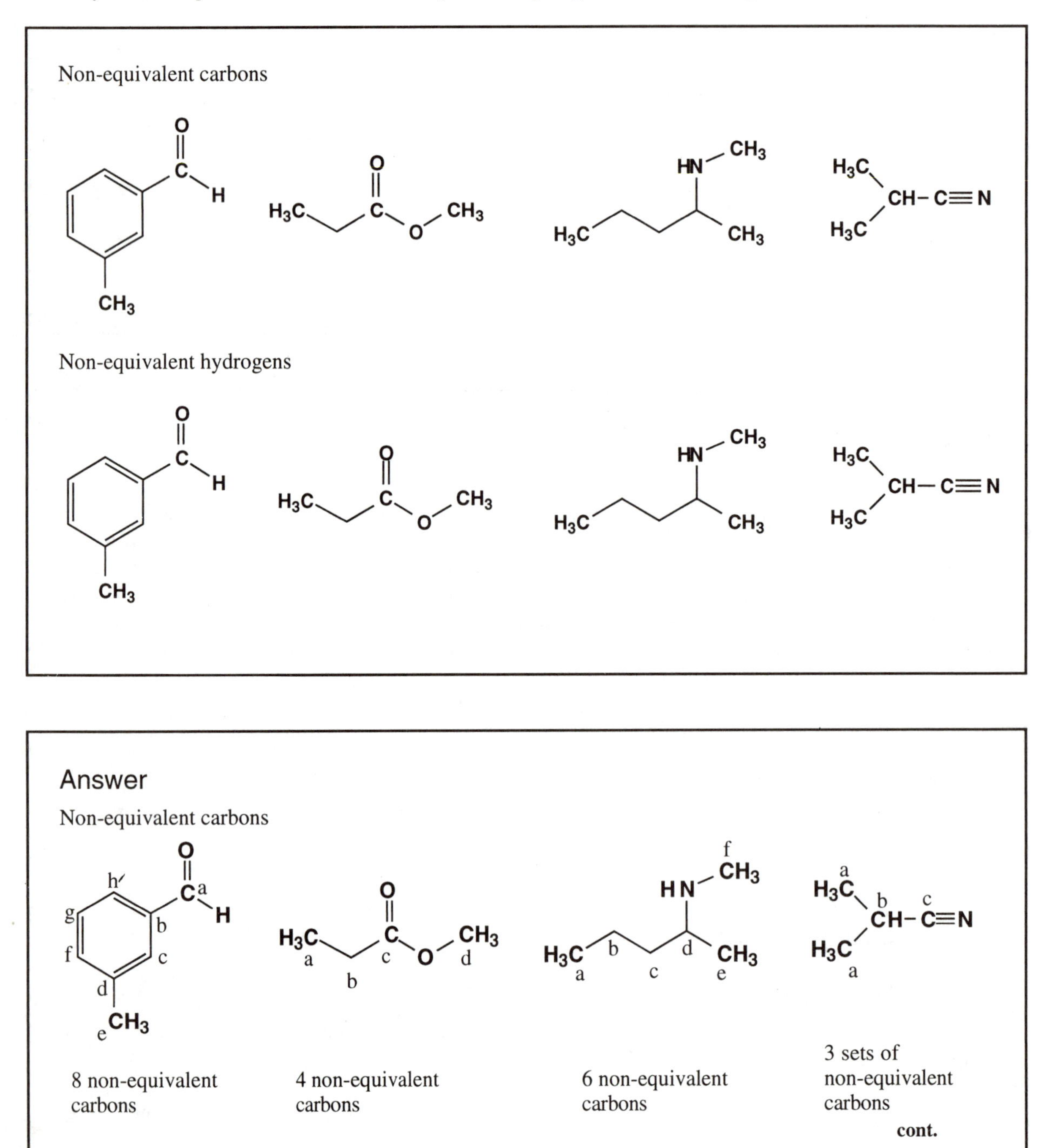

Non-equivalent hydrogens

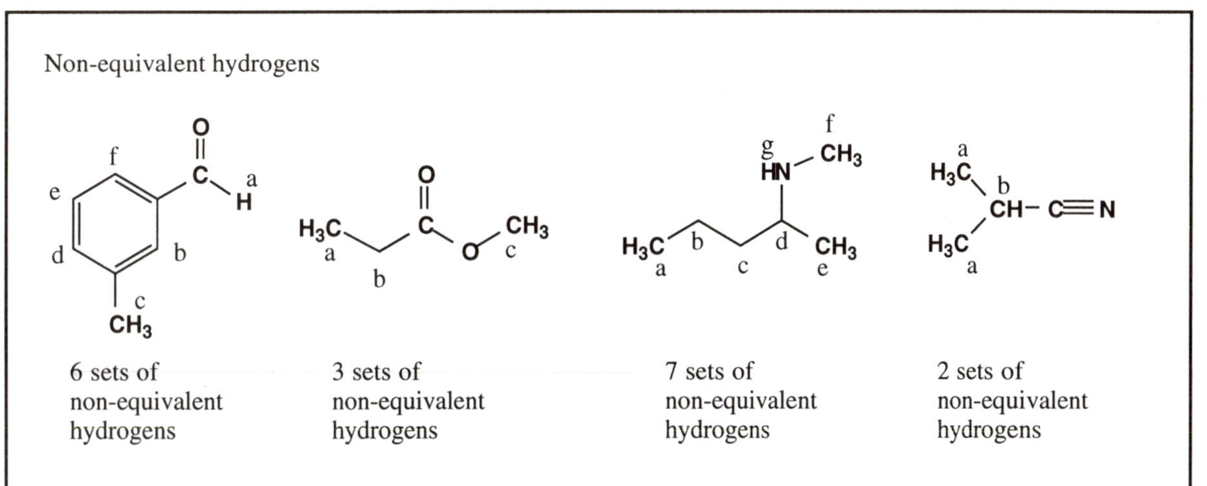

6 sets of non-equivalent hydrogens

3 sets of non-equivalent hydrogens

7 sets of non-equivalent hydrogens

2 sets of non-equivalent hydrogens

When you compare the number of non-equivalent hydrogens to non-equivalent carbons in these structures, the following observations are noted:

• The aromatic compound has two fewer sets of non-equivalent hydrogens since two of the aromatic carbons do not have a hydrogen atom attached.

• The ester has one fewer set of non-equivalent hydrogens since the carbonyl carbon does not have a hydrogen atom attached.

• The amine has one extra set of non-equivalent hydrogens since there is a hydrogen on the nitrogen atom.

• The nitrile has one fewer set of non-equivalent hydrogens since there is no hydrogen on the nitrile carbon.

Summary

- The number of sets of non-equivalent hydrogens in certain molecules will differ from the number of sets of non-equivalent carbons due to the presence of functional groups.
- The following functional groups have carbons without a hydrogen attached:
  ketone, nitrile, carboxylic acid, acid chloride, acid anhydride, ester, amide.
- The following functional groups have a hydrogen present which is not attached to a carbon:
  alcohol, phenol, primary and secondary amines, primary and secondary amides.

SECTION 11

Configurational Isomers—Alkenes

Configurational Isomers—Alkenes

We have looked at constitutional isomers where molecules having the same molecular formula have different molecular structures. These isomers are different compounds having different properties.

We are now going to look at different types of isomers called configurational isomers. These are isomers which *do* have the same molecular structure but which are different because the atoms are arranged differently in space. Furthermore, these different shapes cannot be interconverted without breaking and remaking covalent bonds. As a result, they are different compounds having different properties.

11.1 Alkenes—*cis* and *trans* isomerism

Make a molecular model of 1-butene and describe the structure you obtain. How many 'types' or isomers of 1-butene are possible?

Answer

The alkene portion of the molecule is planar. There is only one possible structure for 1-butene

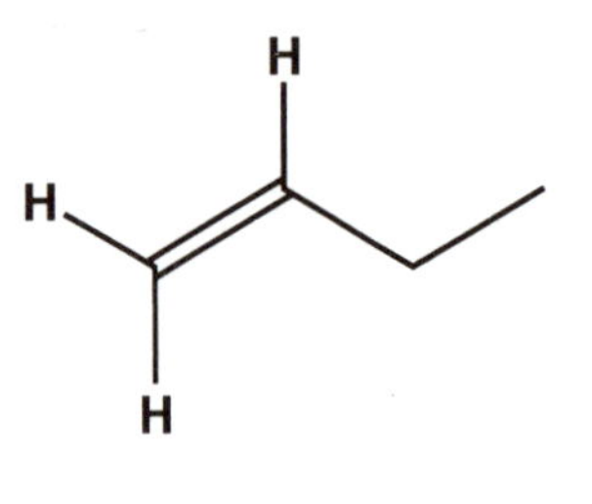

Make models of 2-butene and draw the isomers which are possible.

Answer

There are two possible structures or isomers for 2-butene. In one structure, the methyl groups are on the same side of the double bond (the *cis* isomer). In the other structure, the methyl groups are on opposite sides (the *trans* isomer).

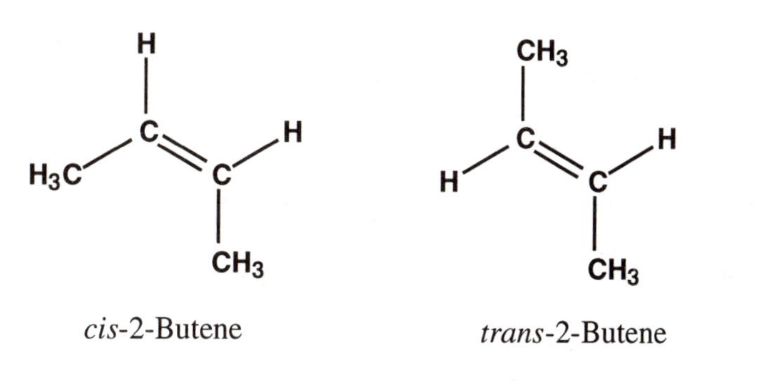

cis-2-Butene

trans-2-Butene

The *cis* and *trans* isomers of an alkene are configurational isomers because they have different shapes and cannot interconvert. Why can they not interconvert?

Answer

The double bond of an alkene cannot rotate and so the substituents are 'fixed' in space. The structures are different compounds with different chemical and physical properties.

The following two structures are different ways of drawing butane. Make a model of butane and decide whether these structures are configurational isomers or not. If not why not?

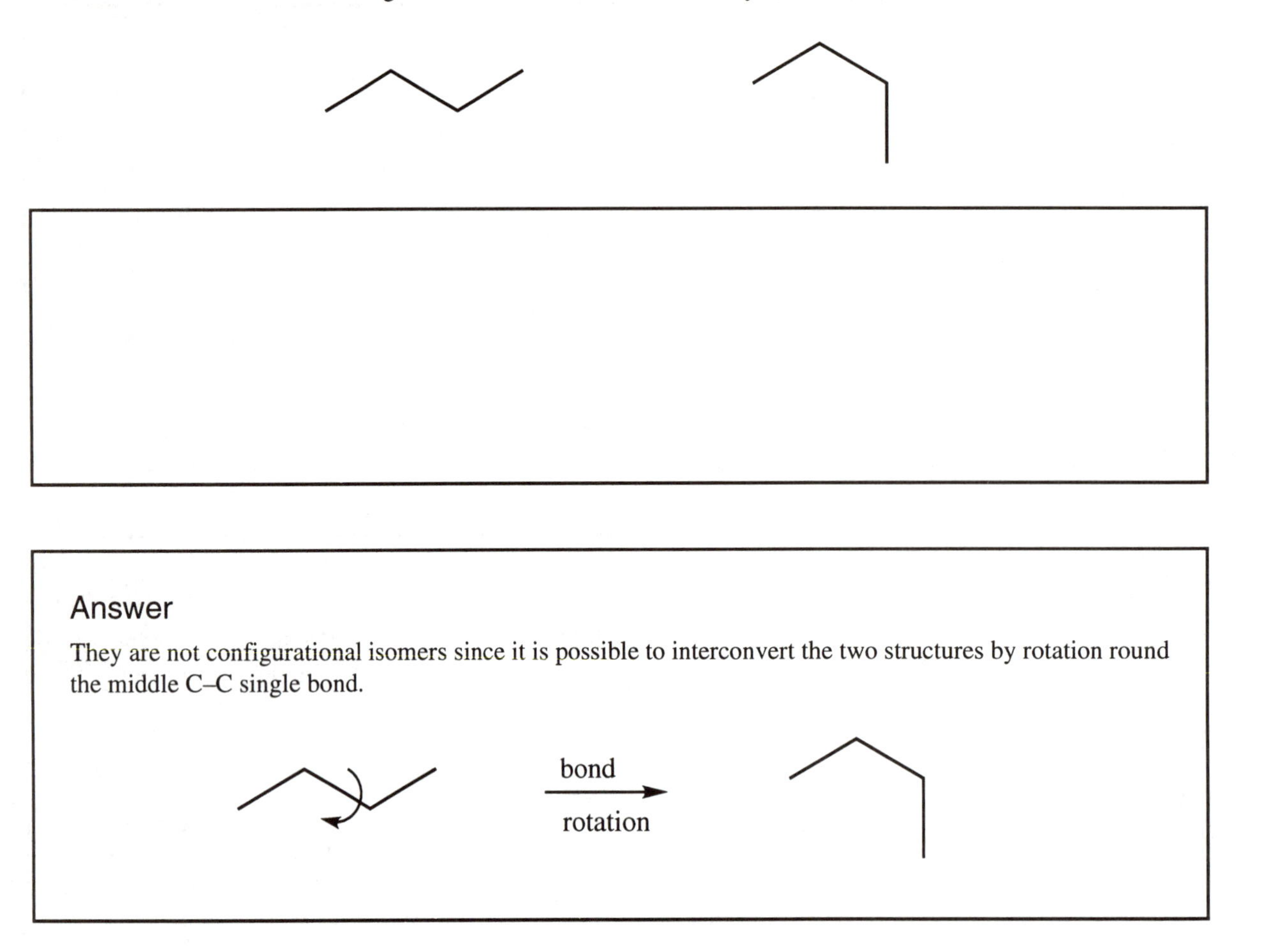

Answer

They are not configurational isomers since it is possible to interconvert the two structures by rotation round the middle C–C single bond.

Draw the *cis* and *trans* isomers of the following alkene and name the isomers.

$$CH_3\sim CH{=}CH\sim CH_2CH_3$$

(Note the wavy line. This is used when you wish to draw an alkene without defining whether it is *cis* or *trans*.)

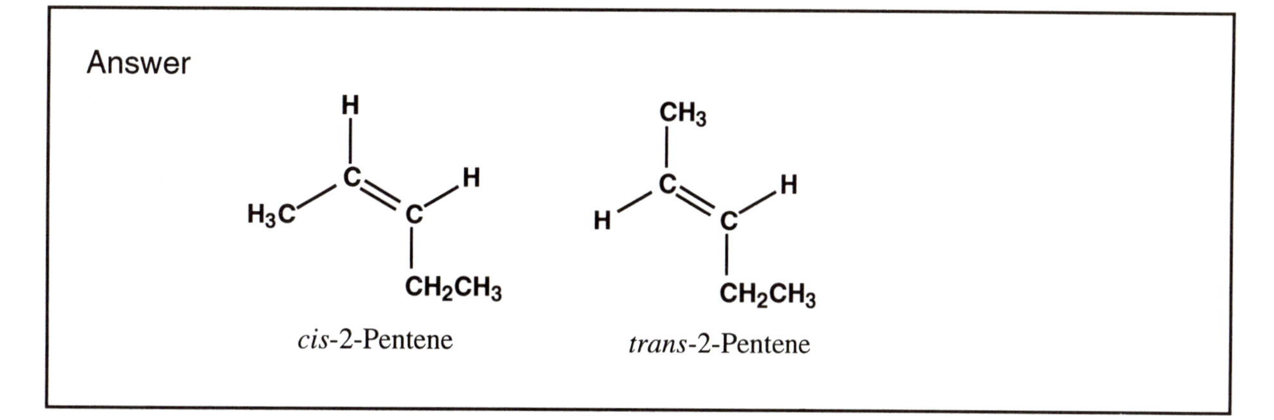

Draw the structure of 2-methyl-1-propene and state whether *cis* and *trans* isomers are possible.

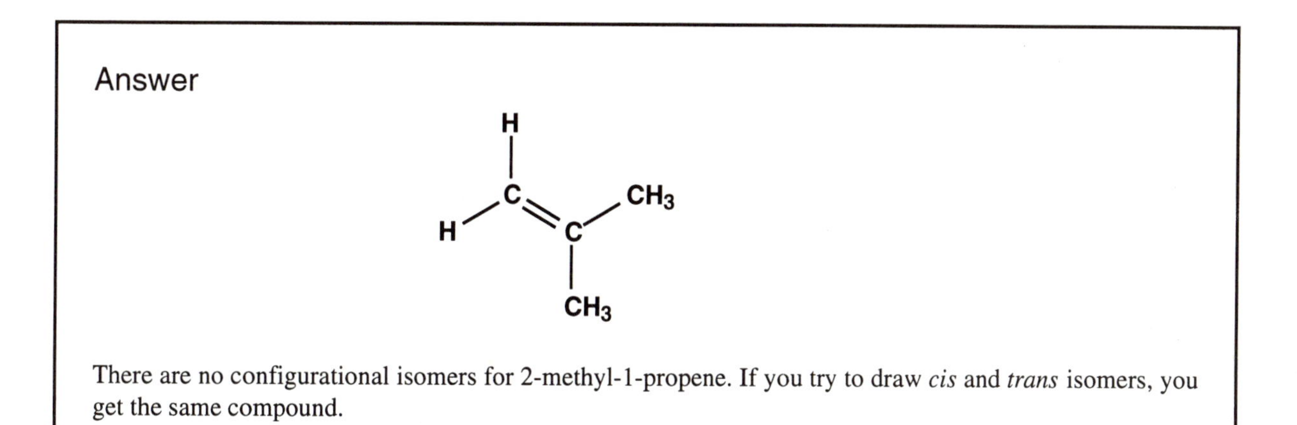

Clearly, not all alkenes can exist as configurational isomers. Draw the following alkenes and decide which can exist as configurational isomers and which cannot.

(a) 2,3-dimethyl-2-pentene, (b) 3,4-dimethyl-3-hexene, (c) 2,3-dimethyl-2-butene, (d) 2-methyl-1-butene.

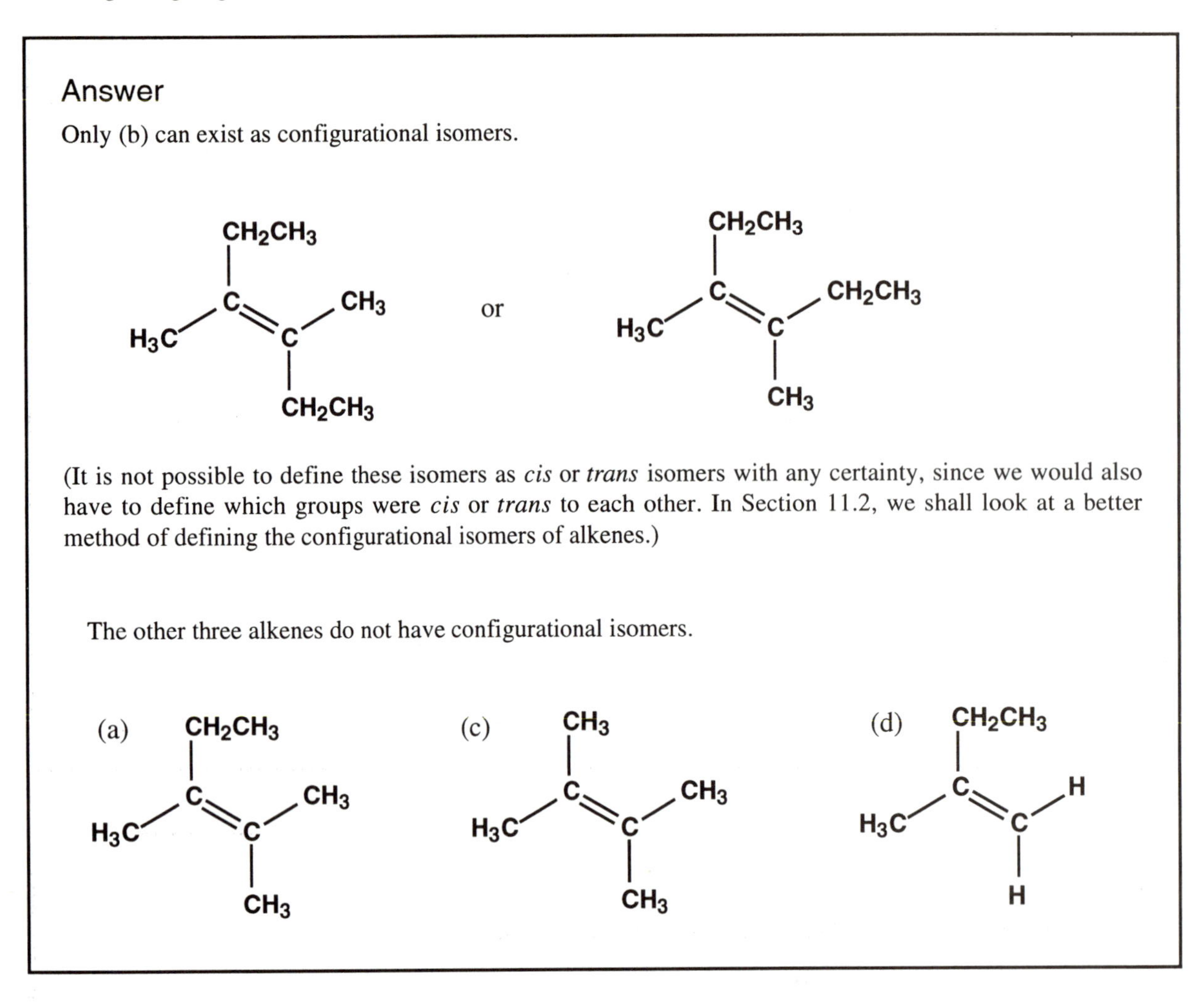

Answer

Only (b) can exist as configurational isomers.

(It is not possible to define these isomers as *cis* or *trans* isomers with any certainty, since we would also have to define which groups were *cis* or *trans* to each other. In Section 11.2, we shall look at a better method of defining the configurational isomers of alkenes.)

The other three alkenes do not have configurational isomers.

Can you come up with a general rule which predicts whether a particular alkene can exist as configurational isomers?

Answer

There must be two different substituents at both ends of the double bond. If there are two identical groups at either end of the double bond then configurational isomers are not possible.

Remember the distinction between constitutional and configurational isomers. The following problem is set to test your understanding of this.

Draw the constitutional and configurational isomers which are possible for the alkene having the formula C_4H_8.

Answers

The functional group is C=C. Possible carbon skeletons are as follows.

Constitutional isomers for skeleton A

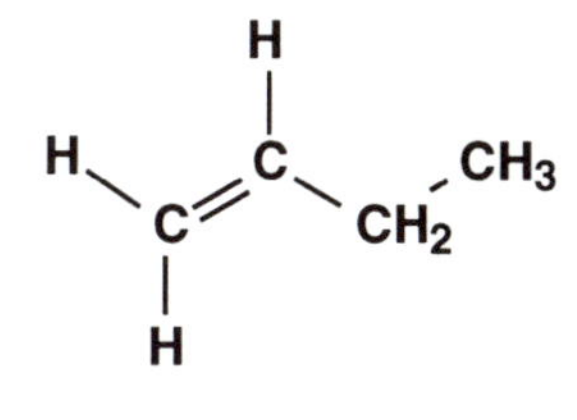

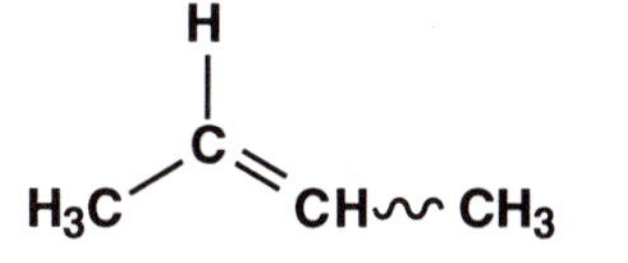

Constitutional isomers for skeleton B

We have identified three constitutional isomers for the alkene formula given. Two of these have the same substituents on at least one end of the double bond and cannot exist as configurational isomers. The third isomer (2-butene) can exist as two configurational isomers and has been drawn with the wavy line to represent this. We now identify the configurational isomers for this alkene.

cont.

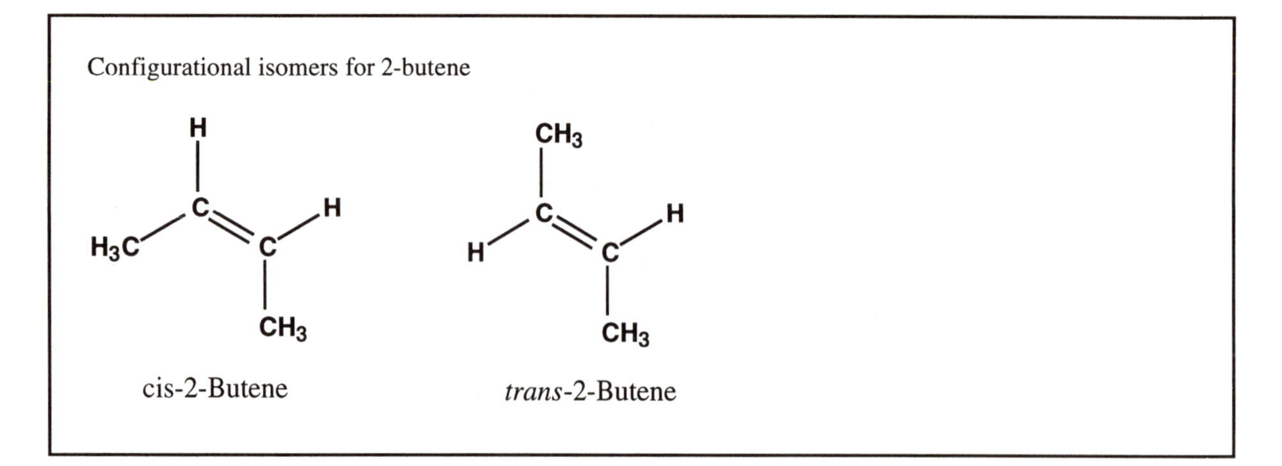

11.2 (*Z*) and (*E*) nomenclature of alkenes

The *cis* and *trans* nomenclature for alkenes is still commonly used, but can be unsatisfactory for some alkenes. Consider the following alkene for example.

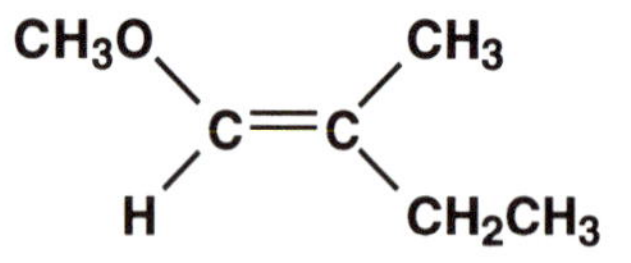

It is difficult to assign this as *cis* or *trans* since this would be ambiguous. Which group is *cis* to which?

The (*Z*)/(*E*) nomenclature allows us to define the configuration of alkenes unambiguously. We shall demonstrate how it works using the above alkene as an example.

- Stage 1. Identify the atoms directly attached to the double bond.

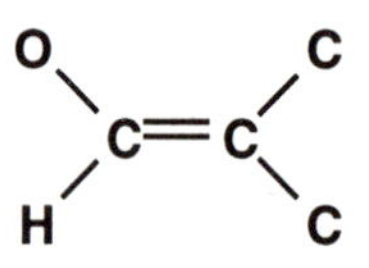

- Stage 2. Write the atomic numbers for each atom.

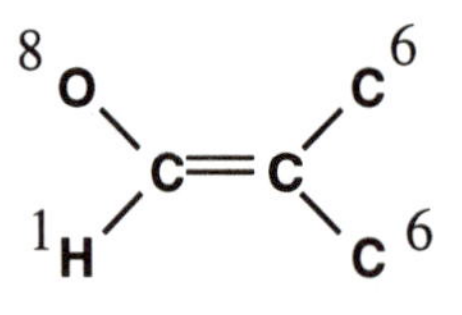

- Stage 3. Compare the two atoms at each end of the alkene. The one with the highest atomic number takes priority over the other.

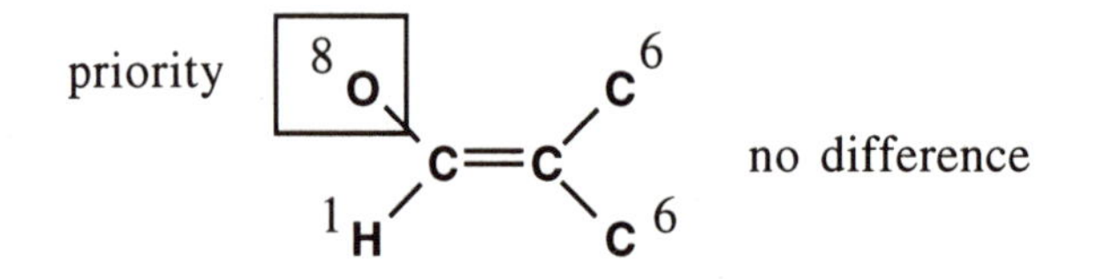

At the left hand-side, oxygen has a higher atomic number than hydrogen and takes priority. At the right hand side, both atoms are the same and we cannot choose between them. We need to move to Stage 4.

- Stage 4. If there are identical atoms at either end of the double bond, compare the atoms of highest atomic number attached to them.

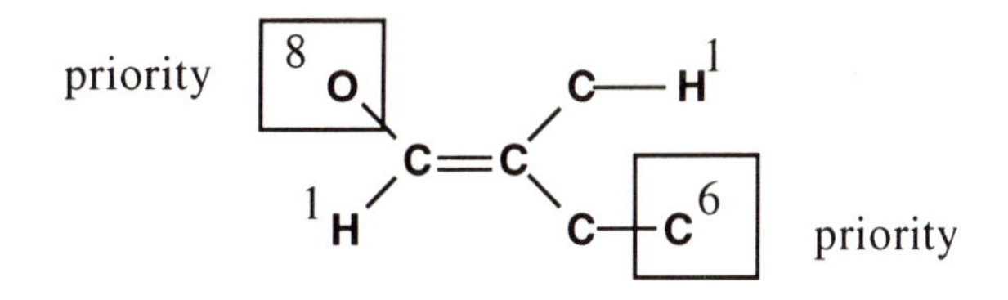

- Stage 5. If the two priority groups are on the same side of the double bond, the alkene is designated as (*Z*) (from the German word 'zusammen' meaning together).

  If the two priority groups are on opposite sides of the double bond, the alkene is designated as (*E*) (from the German word 'entgegen' meaning across).

  In this example, the priority groups are on opposite sides and so the alkene is (*E*).

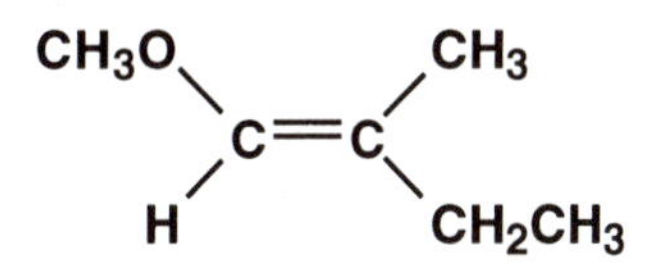

(*E*)-1-Methoxy-2-methyl-1-butene

Assign the following alkene as (*E*) or (*Z*).

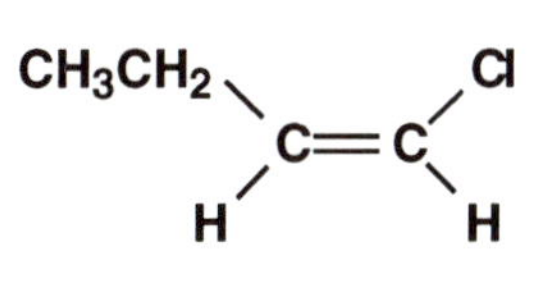

Answer

This alkene is (*Z*)-1-chloro-1-butene. The groups which take priority at each end of the alkene are on the same side of the alkene.

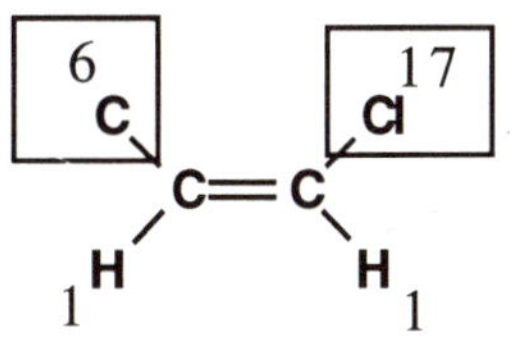

Summary

- Configurational isomers have the same molecular formula and the same bonding framework, but the atoms are arranged differently in space. The isomers cannot be interconverted without breaking bonds.
- Alkenes cannot rotate around the double bond and can exist as configurational isomers.
- An alkene having identical substituents at either end of the double bond cannot have configurational isomers.
- The configuration of alkenes is defined as being (*Z*) or (*E*) depending on whether the groups of highest priority are on the same side or opposite sides of the alkene.

SECTION 12

Configurational Isomers—Optical Isomers

Configurational Isomers—Optical Isomers

The *cis/trans* or (E / Z) isomerism of alkenes is one example of configurational isomerism. There is another example of configurational isomerism known as optical isomerism, so named because of the ability of optical isomers to rotate plane polarised light clockwise or anticlockwise. At first sight, this may not appear to be particularly important or exciting, but the existence of these optical isomers has very important consequences for life. You may have heard the statement that life is 'left-handed'. The reason for this will become clear once you understand what optical isomers are.

In order to understand why compounds can be optical isomers, you need to appreciate the shape of molecules and it is crucial that you use molecular models to assist you in this section.

12.1 Asymmetric molecules

Construct molecular models of the following molecules. Describe the shapes and identify whether it is possible to construct a different structure for any of these molecules:

CCl_4 $CHCl_3$ CH_2Cl_2 $CH_2Cl(OH)$

Answer

All these molecules are tetrahedral and there is only one way of fitting the atoms together.

Now construct two models of lactic acid and compare them. Is there any way you can construct two models of lactic acid such that they are different?

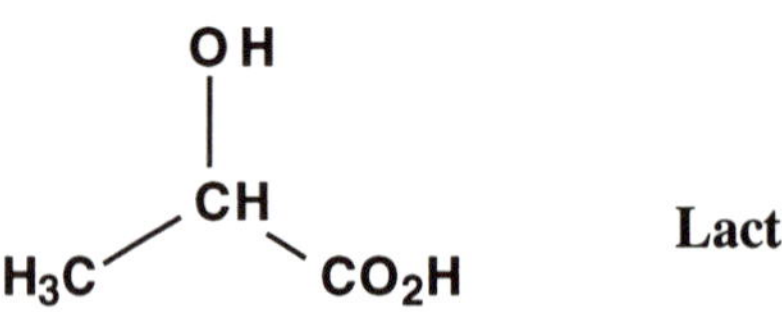

Lactic acid

Answer

There are two different ways of making the model of lactic acid. If there is no difference between the two models you have made, try swapping round the OH group and the H atom attached to the central carbon of one of the models.

You should now have two different models which cannot be superimposed on each other. Since they cannot be superimposed on each other, they must be isomers. Identify the difference between them and draw the structures to show the difference.

Answer

The difference between the two possible models is the way the substituents are attached to the central carbon. This can be represented by the following drawings, where the bond to the hydroxyl group is coming out the page in one isomer but going into the page in the other isomer.

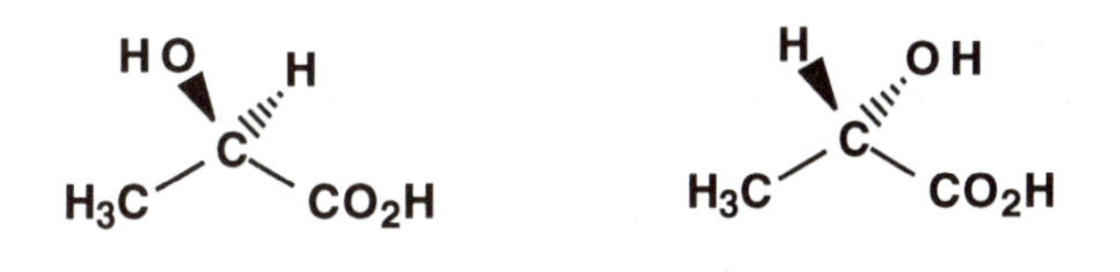

The two isomers of lactic acid are related in some way. How? Compare the two models if you are not sure.

Answer

They are non-superimposable mirror images of each other. This can be demonstrated by comparing the models as below

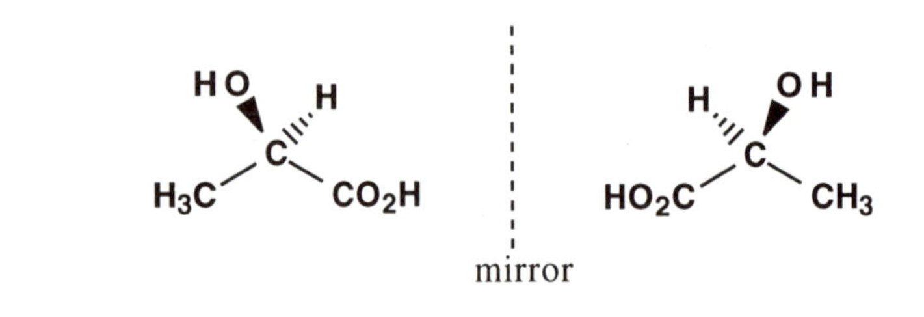

Take your models of the two possible lactic acids and convert them to the following structure I by replacing the hydrogen attached to the central carbon with another CH_3 group i.e.

I 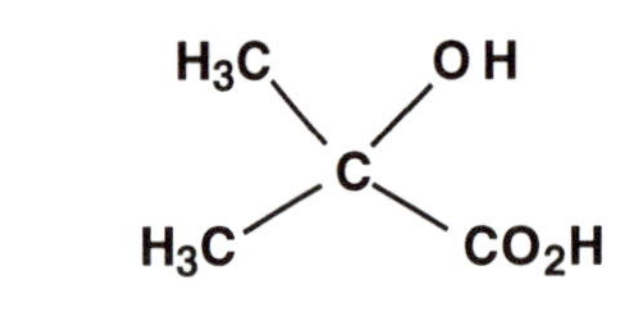

Are the two models still different?

Answer

The two models are still mirror images but they are now superimposable and are therefore identical. They are no longer configurational isomers.

Why should there be this difference?

Compare the symmetries of lactic acid and structure I. Can you identify any planes or axes of symmetry?

Answer

Lactic acid has no elements of symmetry whereas structure I has a plane of symmetry.

Unsymmetrical molecules which lack any planes or axes of symmetry (e.g. lactic acid) are described as being 'chiral' or 'asymmetric'. They can exist as two non-superimposable mirror images. These mirror image structures are known as enantiomers and are another example of configurational isomer. (Molecules containing only one axis of symmetry and no planes of symmetry are also chiral and are discussed later in this section.)

12.2 Asymmetric centres

It is not always easy to identify whether a molecule is chiral by identifying whether it has any elements of symmetry. Fortunately, there is one simple method which involves identifying what are known as asymmetric centres. This works for most chiral molecules, but it is important to realise that it is not foolproof and that there are several cases where it will not work, e.g:

(a) some chiral molecules have no asymmetric centres;

(b) some molecules having more than one asymmetric centre are not chiral.

Some examples are shown in Section 12.9, but for the moment we shall concentrate on molecules which do not have these complications.

Taking the last exercise into account, can you come up with a simple rule to predict whether a molecule will have optical isomers or not?

Answer

If a compound is to have optical isomers, there must be four different substituents attached to a central carbon.

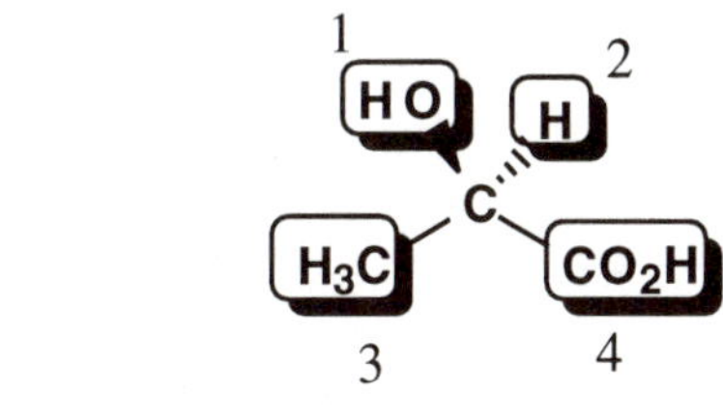

Whenever there are four different substituents attached to a single carbon, the mirror images are non-superimposable and the structure can exist as two configurational isomers. The carbon centre which contains these four different substituents is known as a **stereogenic** or **asymmetric centre**.

Such configurational isomers are known as **enantiomers** or **optical isomers**. That is because a solution of each isomer is capable of rotating plane polarised light. One isomer will rotate plane polarised light to the left while the other (the mirror image) rotates it to the right by the same amount. An equal mixture of the two isomers (a **racemate**) will not rotate plane polarised light at all. In all other respects (bar one), the two isomers are identical in physical and chemical properties and are therefore indistinguishable. There is one important exception to this as we shall see later, and it is crucial to the functioning of life.

It is possible to predict whether molecules can exist as a pair of enantiomers by looking at the symmetry (or lack of it) within the molecule, but it is usually easier to identify an asymmetric centre. If you spot an asymmetric centre, the molecule will usually be chiral or asymmetric and can exist as a pair of enantiomers. (There are exceptions which we shall consider in Section 12.9.)

Identify any asymmetric centres in each of the following and hence predict whether each molecule is chiral or achiral (not chiral).

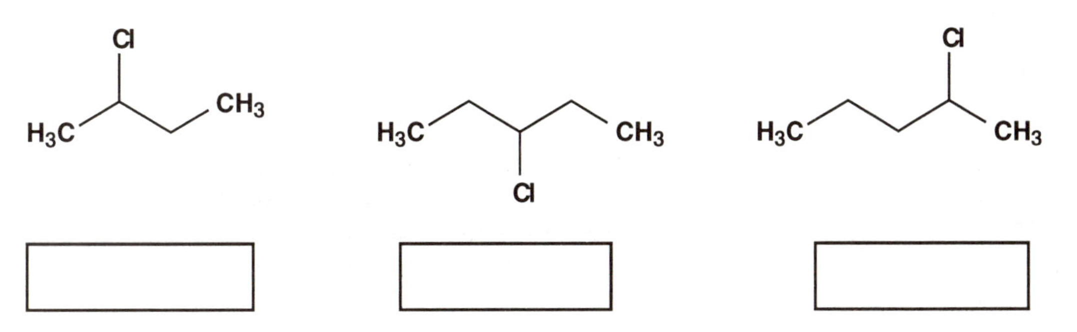

Answer

The first and the third molecules have an asymmetric centre marked (*). They are therefore chiral molecules and can exist as two enantiomers or optical isomers.

The second structure has no asymmetric centre and is symmetric or achiral. It does not have optical isomers.

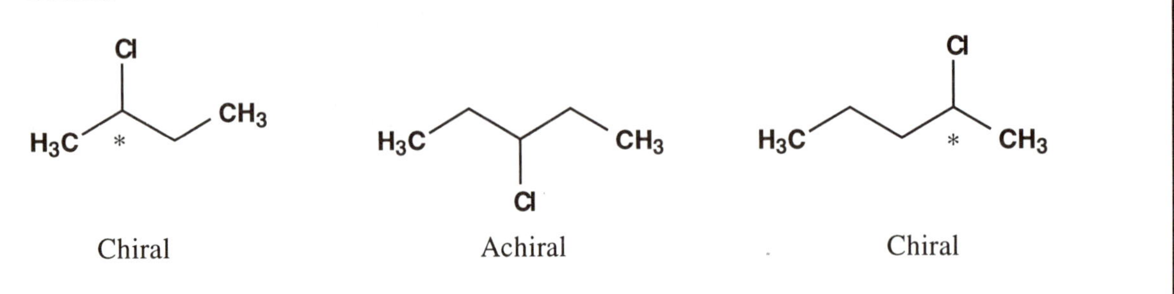

A molecule can have more than one asymmetric centre. Identify the asymmetric centres in the following structures.

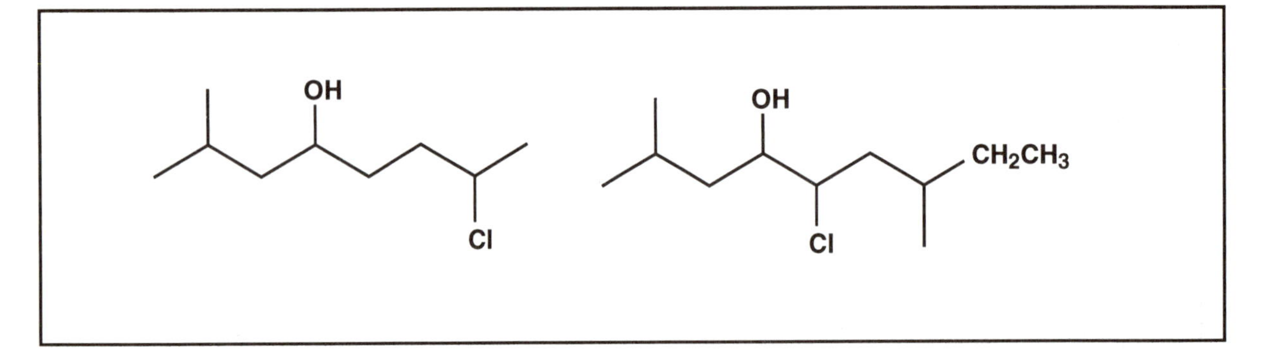

Answer

Remember an asymmetric centre has to have four different substituents and so you can ignore any carbon atoms such as CH_3 or CH_2. You only need to consider tertiary or quaternary carbons as possible asymmetric centres.

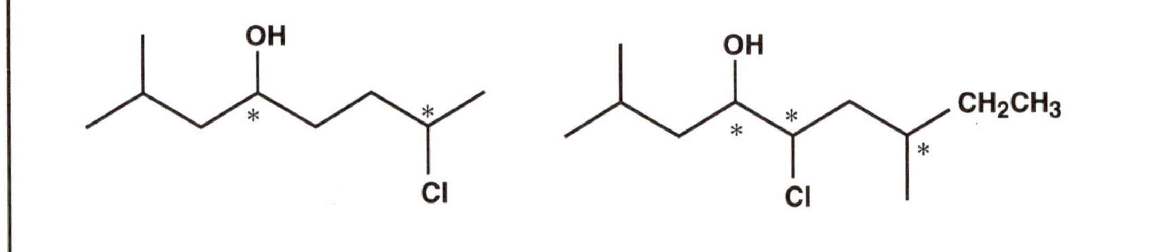

12.3 Asymmetric centres in cyclic compounds

Identifying asymmetric centres in cyclic structures can sometimes be difficult. Carry out the following procedure.

- Identify tertiary and quaternary carbon atoms in the ring.
- Consider both halves of the ring attached to the carbon centre. If there is any difference between the two halves then this is equivalent to two different substituents and the centre is asymmetric.

Example

Is the following cyclic structure chiral? If so, identify any asymmetric centres.

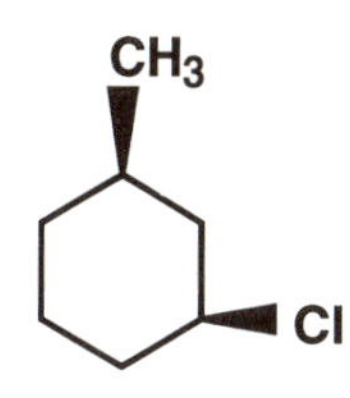

Method

(1) Identify any tertiary or quaternary carbons in the ring.
In this case, there are two possible carbons marked as shown.

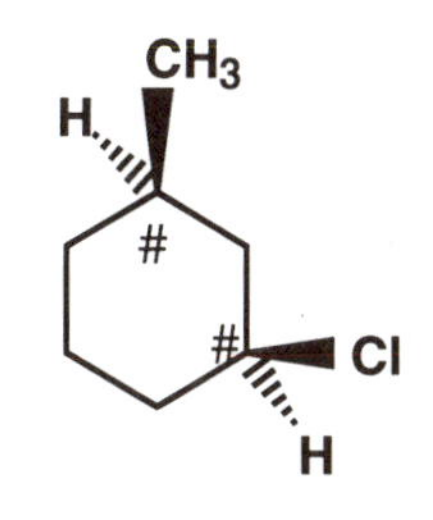

(2) Consider the top carbon and compare each half of the ring to which it is attached. One half is different from the other since there is a chlorine atom attached to one side but not the other. The top carbon is therefore an asymmetric centre.

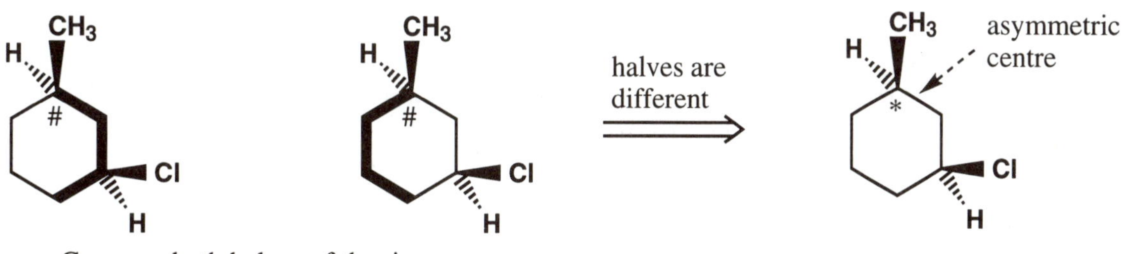

Compare both halves of the ring

(3) Repeat the exercise for the carbon containing the chlorine atom. This, too, is an asymmetric centre.

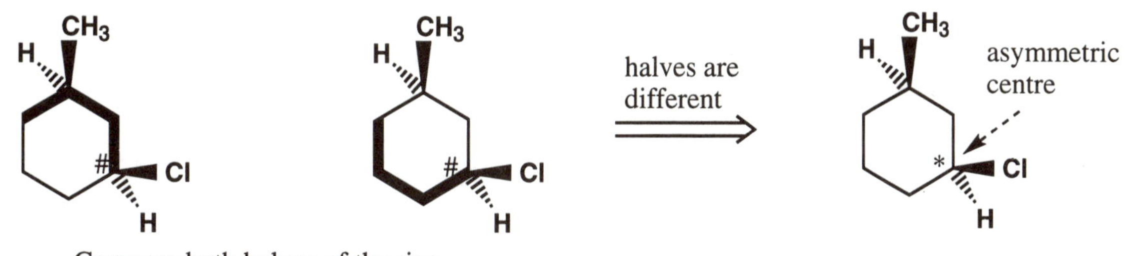

Compare both halves of the ring

(4) There are therefore two asymmetric centres in the molecule.

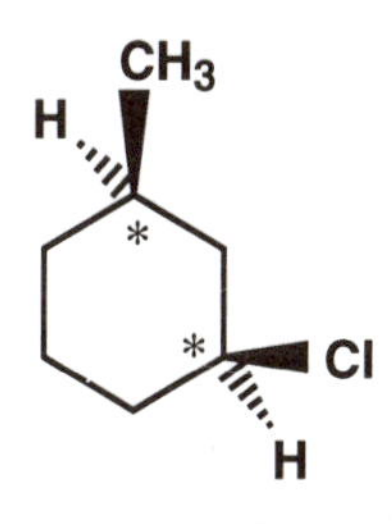

Drawing all these structures out as above can get laborious and, with practice, you should be able to carry out the procedure in your head. Try the following two structures and identify any asymmetric centres which may be present.

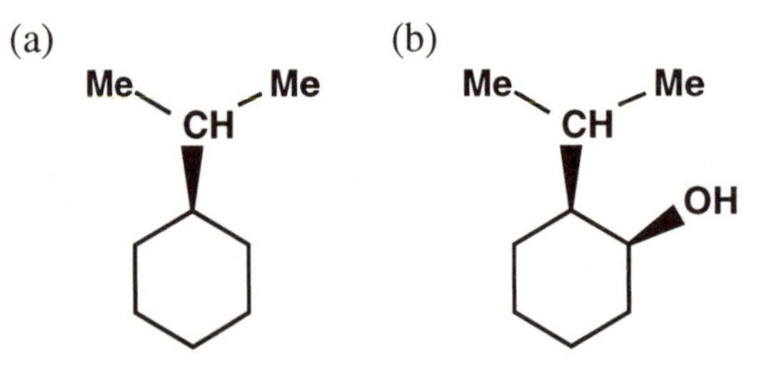

Answer

Structure (a) has no asymmetric centre. There is a tertiary carbon in the ring but both sides of the ring are the same. The molecule has a plane of symmetry and as a result cannot be chiral. Molecule (b), on the other hand, has two asymmetric centres.

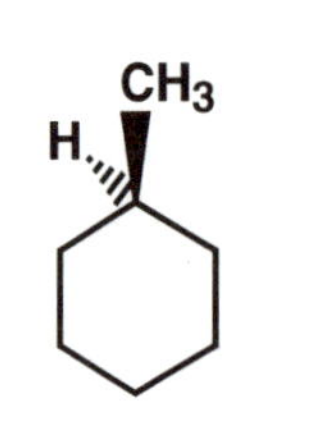

no asymmetric centres

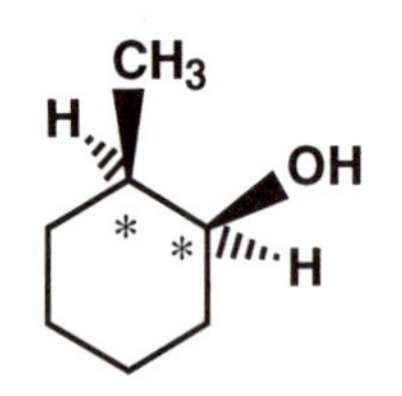

two asymmetric centres

One of the following molecules is chiral but the other is not. Identify the chiral molecule and the asymmetric centres within it.

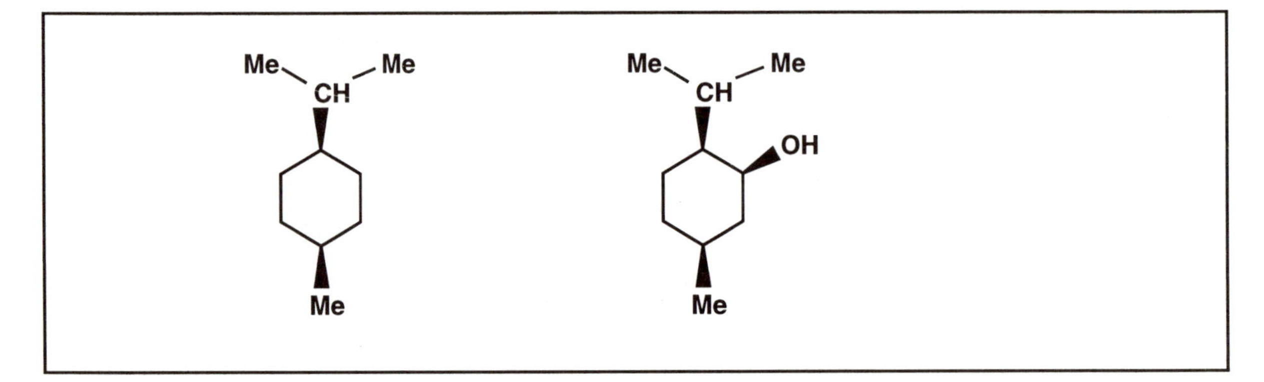

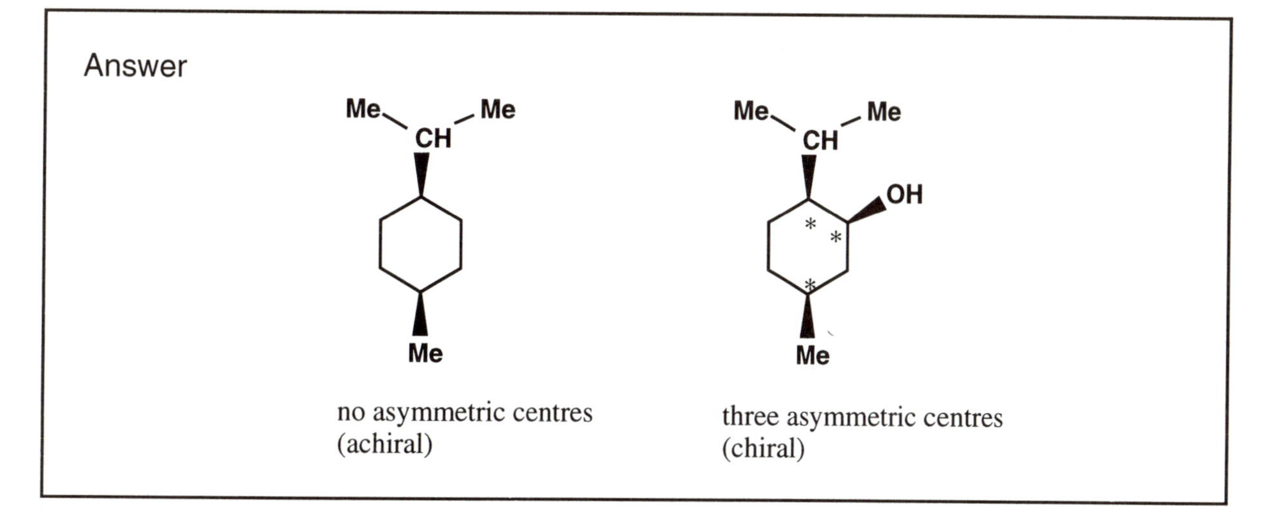

If a molecule is achiral (not chiral), then there must be some element of symmetry present which prevents its mirror image being non superimposable. Can you identify any such element of symmetry in the achiral molecule below?

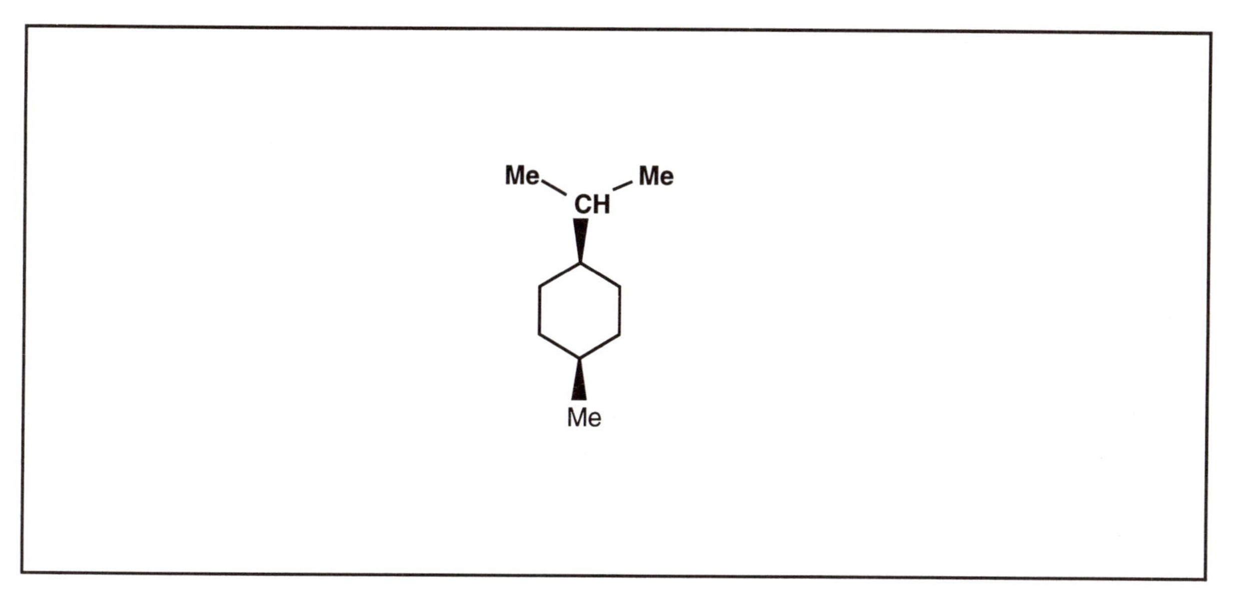

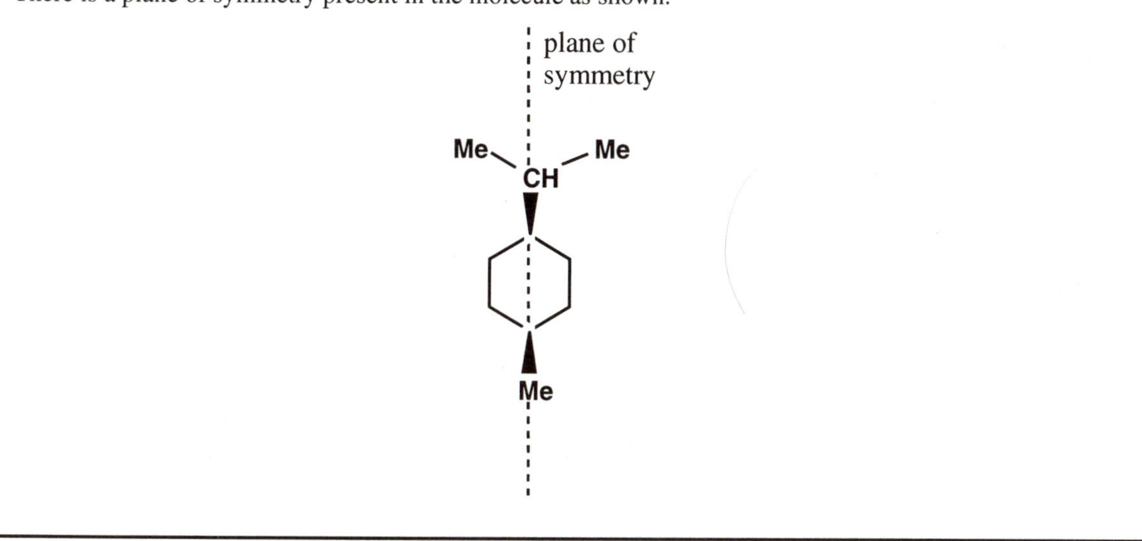

12.4 Symmetry of sp^2 and sp centres

Asymmetric centres are only possible on sp^3 carbons. An sp^2 or an sp centre cannot be an asymmetric centre. Why not? (Remember that chirality is a lack of symmetry.)

Consider the following alkene and alkyne in your answer and compare the mirror images of both structures.

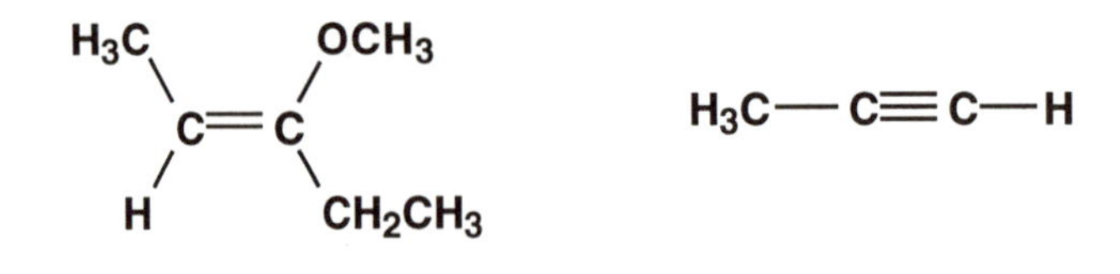

Answer

An sp^2 centre in an alkene or a carbonyl group is planar and cannot be asymmetric, i.e. the mirror images are superimposable.

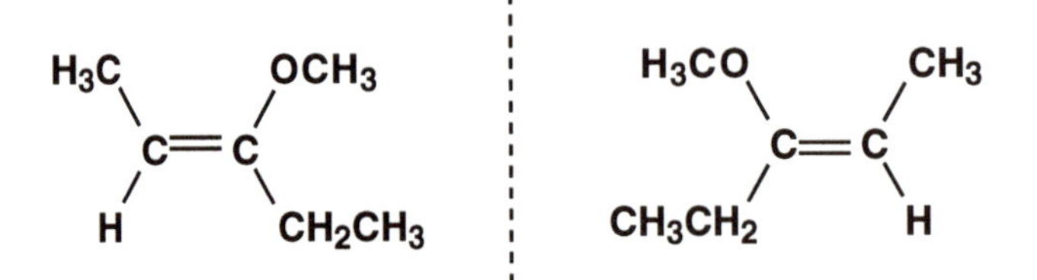

Superimposable mirror images
(i.e. identical molecules)

Similarly, the bonds of an sp centre are linear with respect to each other and cannot give an asymmetric shape.

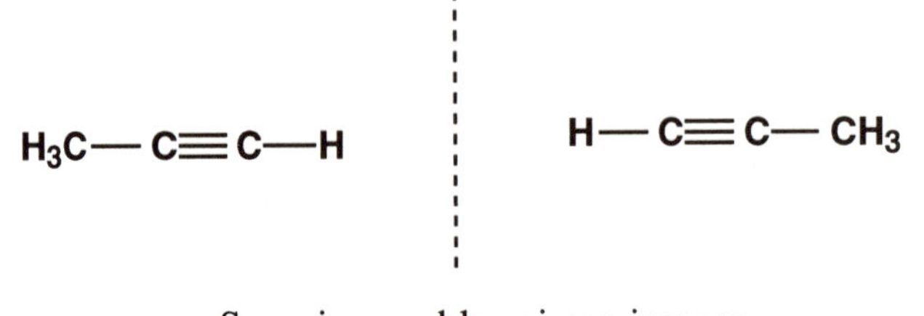

Superimposable mirror images
(i.e. identical molecules)

12.5 Identifying asymmetric centres in more complicated molecules

With complicated molecules, it takes a long time to look at each individual carbon and to count the substituents. You may find it easier to quickly eliminate the carbons which cannot possibly be asymmetric as a first step.

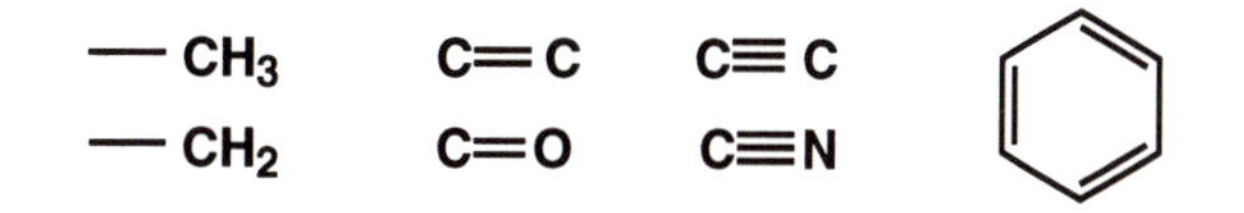

Once you have ruled these out, you can concentrate on the remaining carbons and decide whether they are asymmetric centres or not.

Noradrenaline is an important neurotransmitter in the body, responsible for passing messages between nerves. Identify any asymmetric centres and draw the possible optical isomers.

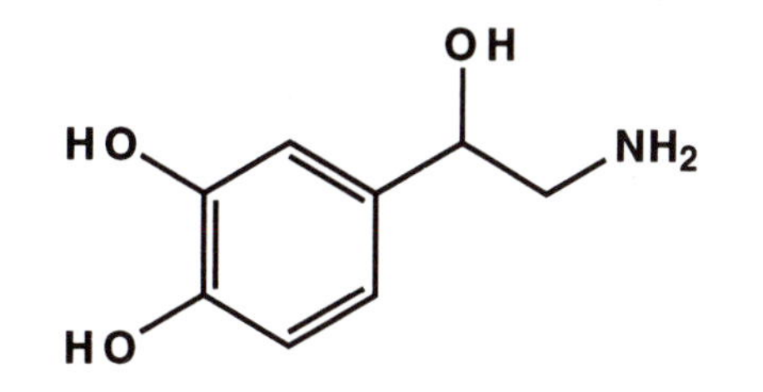

Answer

The aromatic carbons can be ruled out, as can the CH_2 group. This leaves only one carbon and this does have four different substituents. Therefore, there is one asymmetric centre in noradrenaline (marked with an asterisk). The molecule is chiral and can therefore exist as two non superimposable mirror image forms. Therefore, there are two optical isomers (enantiomers) possible.

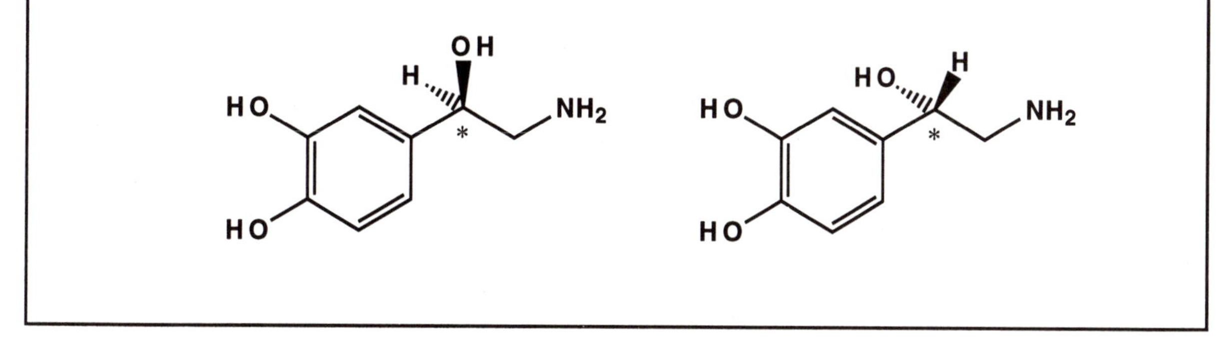

Only one of these enantiomers is present in the body.

12.6 The chirality of life

The following molecules are important molecules in cell biochemistry. Identify any asymmetric centres and hence identify which molecules can exist as a pair of enantiomers.

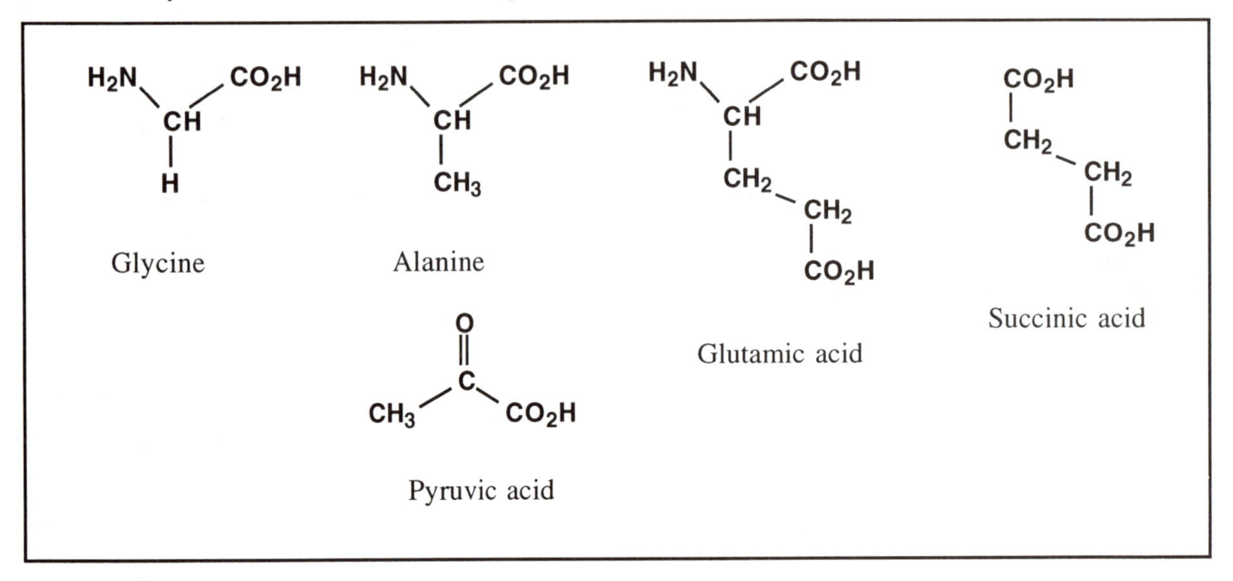

Answer

The asymmetric centres are marked with an asterisk. Those molecules with asymmetric centres can exist as enantiomers or optical isomers.

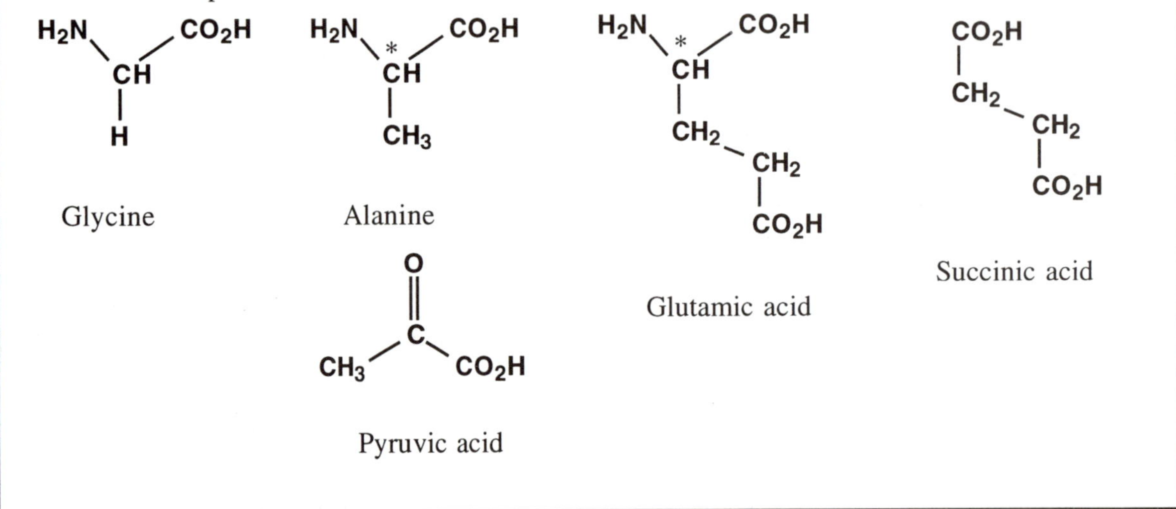

The two structures above which are chiral (alanine and glutamic acid) are examples of naturally occurring amino acids. There are two possible optical isomers for each amino acid but only one of them exists naturally in the human body and this is known as the L-form. Since enantiomers are non-superimposable mirror images of each other, the expression 'handedness' is sometimes used for this relationship. A left hand is a non-superimposable image of a right hand. Since only one enantiomer of amino acids is naturally occurring, it is the equivalent of having only one hand—in the case of amino acids, the left hand.

Thus, amino acids and life can be described as being left handed.

Enzymes are protein molecules which act as the cell's catalysts and are made up of amino acids. Since amino acids are chiral, an enzyme is also chiral and can exist as two non-superimposable mirror images. However, only one of these mirror image forms exists naturally. Why?

Answer

The building blocks for enzymes are amino acids. Since only one of the enantiomers is present naturally for each amino acid, they will construct only one of the possible mirror image forms of the enzyme.

The fact that enzymes are chiral and only exist as one form has important consequences for life. Enzymes help to catalyse reactions within the cell and one of the ways they do this is to temporarily bind reagents before they react. This binding process allows enzymes to distinguish between the two possible enantiomers of a chiral compound.

Consider the following reaction where lactic acid is converted to pyruvic acid with the aid of an enzyme called lactate dehydrogenase.

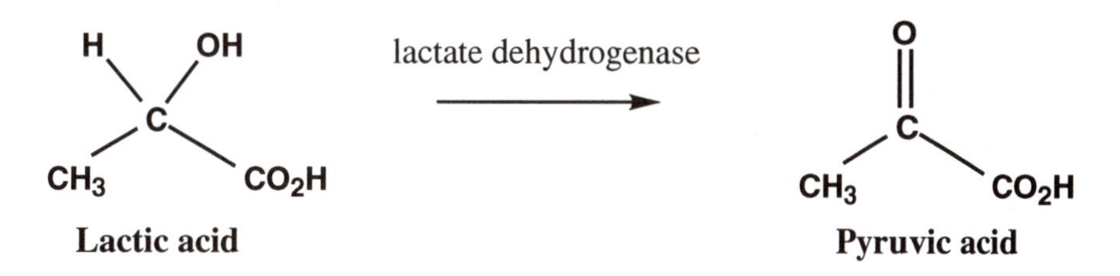

Only one of these molecules is chiral. Identify which one and draw both possible enantiomers.

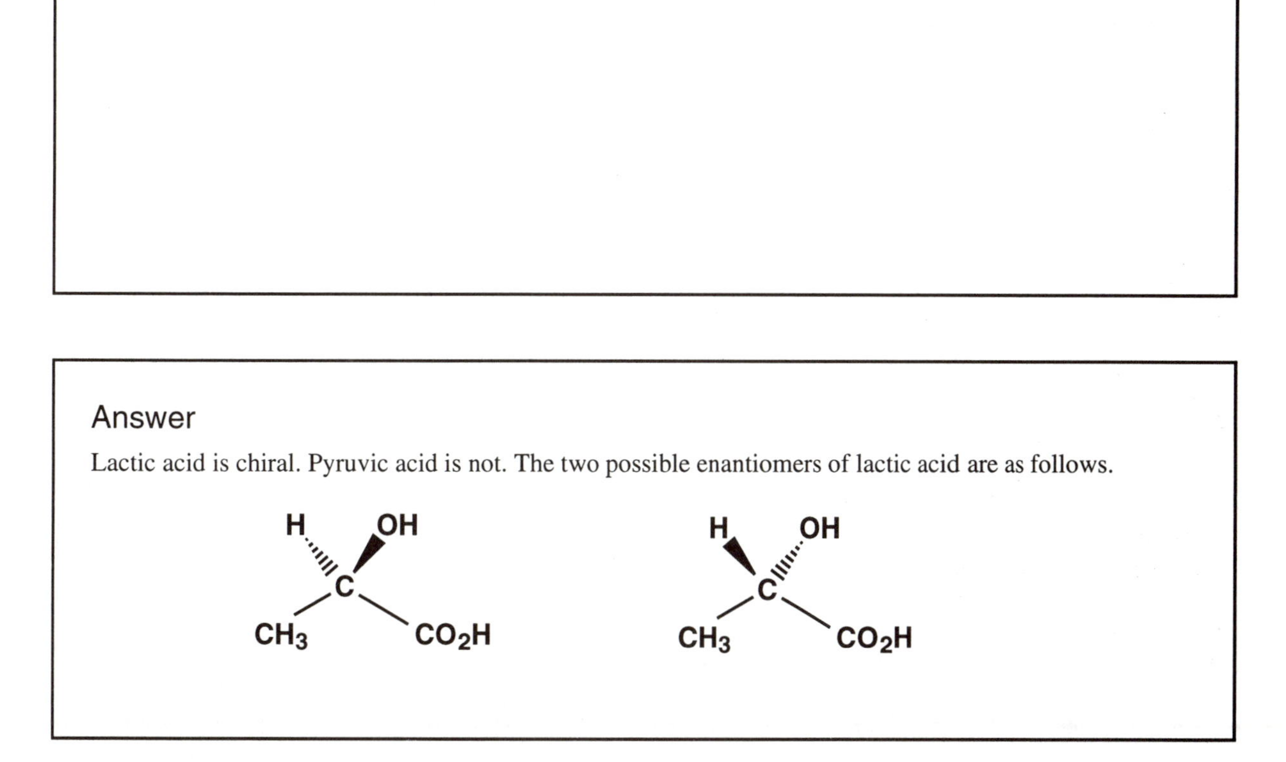

Answer

Lactic acid is chiral. Pyruvic acid is not. The two possible enantiomers of lactic acid are as follows.

Only one of these enantiomers takes part in the enzymic reaction. What does this imply in terms of the enzyme's ability to bind each of these enantiomers?

Answer

It implies that the enzyme is capable of binding one of the enantiomers, but not the other. Only the enantiomer which binds to the enzyme reacts.

Let us assume that the enzyme has three binding sites responsible for binding the methyl, the acid and the hydroxyl groups of lactic acid, as shown in the diagram below. Are both enantiomers of lactic acid capable of binding to these sites using the correct groups?

If you have trouble visualising this, draw the three binding sites on a piece of paper, then take the molecular models you have built for the two enantiomers of lactic acid and try to fit the relevant groups to their binding sites. Draw how the two enantiomers interact with the enzyme in the box below.

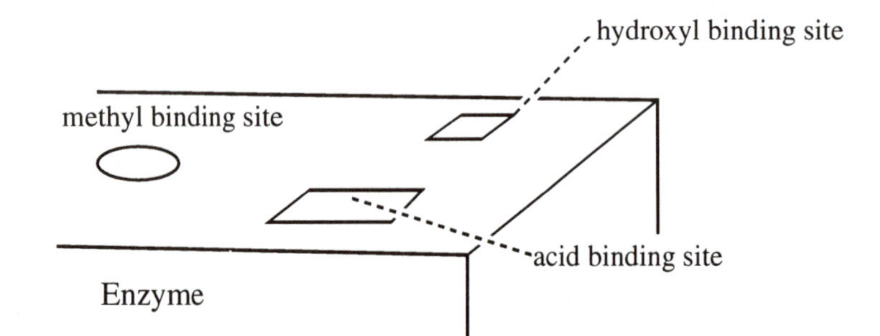

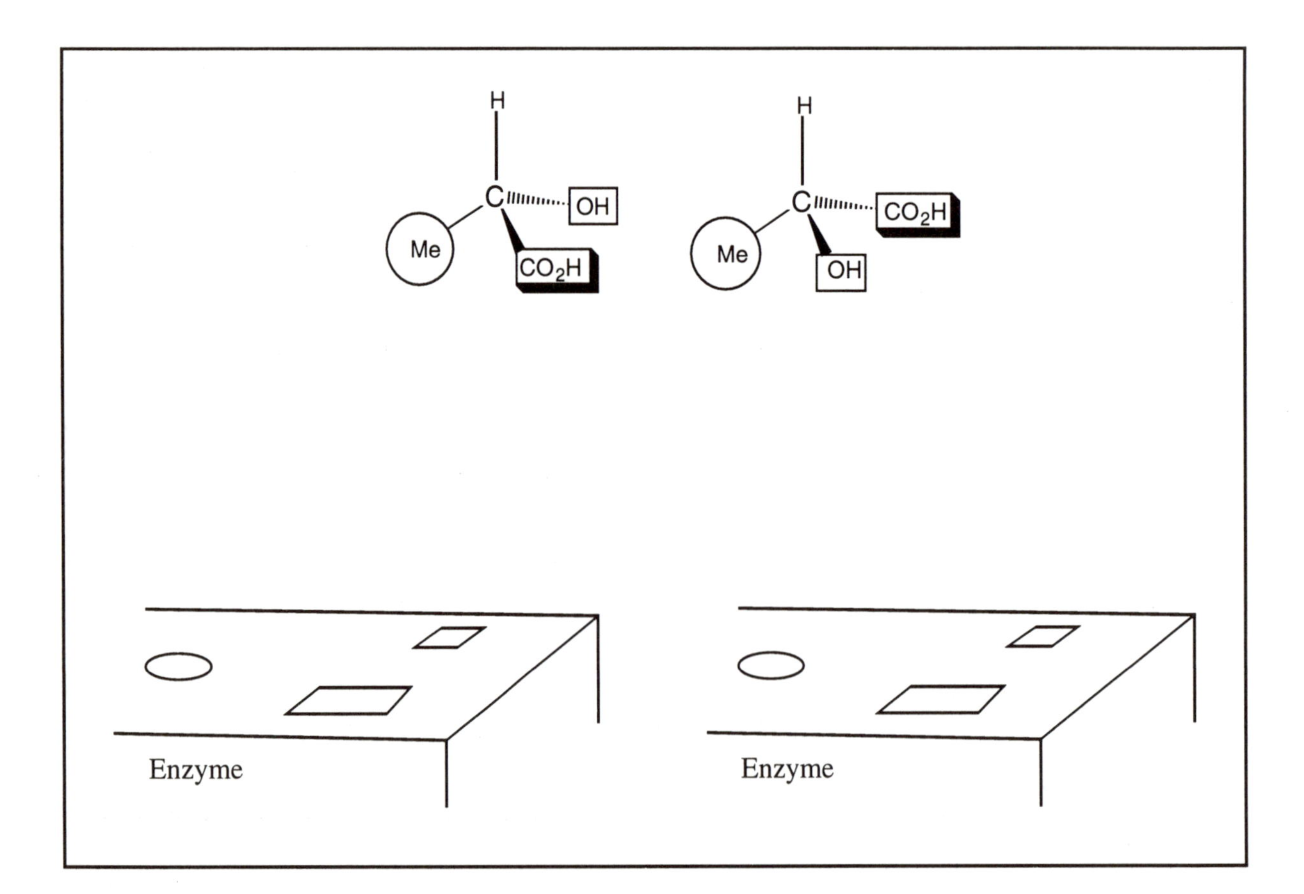

Answer

From the following diagrams, you can see that one enantiomer can be bound at its three binding sites whereas the other cannot. The enzyme therefore ignores one of the enantiomers and only one enantiomer takes part in the reaction.

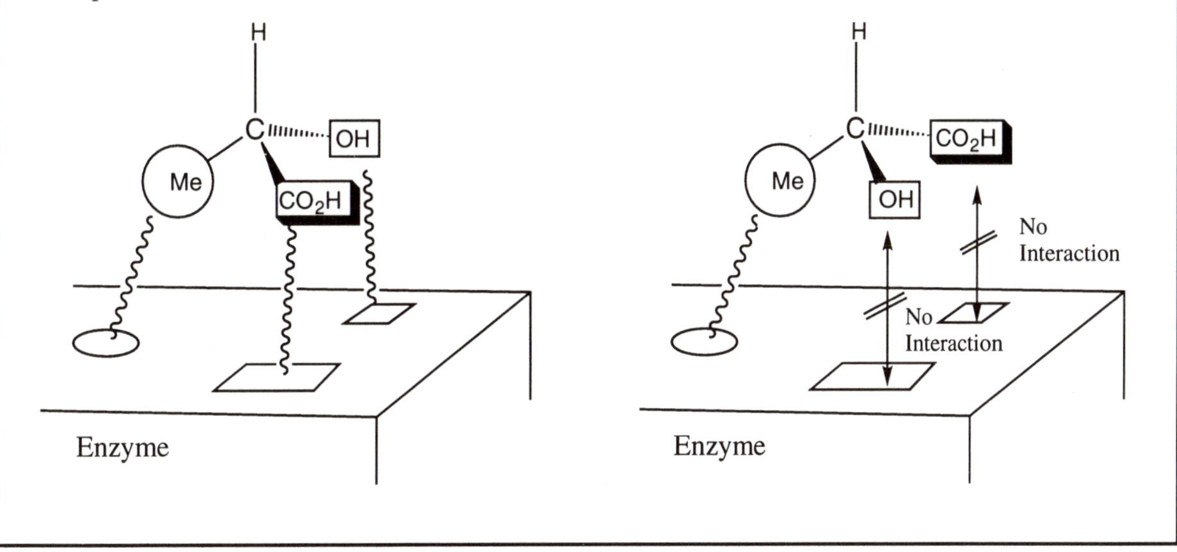

Enzymes are crucial to the catalysis of most reactions which proceed in the human body. Since they are chiral and only exist as one form, they will recognise only one enantiomer of a chiral compound. If the reaction they catalyse generates a chiral compound, then only one enantiomer is produced. For example, lactate dehydrogenase can also catalyse the reverse reaction where pyruvic acid is converted to lactic acid. Only one enantiomer of lactic acid is formed.

The presence of enzymes helps to explain why chiral molecules only exist as single enantiomers in the body today, but it does not explain how this preference originated. Enzymes catalyse the formation of optically active amino acids because they are made up of optically active amino acids. But how did the L-enantiomers of amino acids get preferred in the first place?

If we are going to tackle that question, we have to look at how life began on the planet.

Scientists have being tackling this problem for many years. It is generally thought that the first amino acids to appear on earth were formed in an atmosphere of methane, ammonia and hydrogen sulphide. Various laboratory experiments have been carried out to see if amino acids can be synthesised under these conditions. Under the influence of an electric discharge (lightning), it has been found that amino acids can indeed be formed. However, the amino acids are a 50/50 mixture of both possible enantiomers.

If life only uses one enantiomer, what happened to the other one?

Answer

This is one of the mysteries of science and nobody can be sure of the process by which one enantiomer was selected over the other for life's building blocks.

It might just have been a chance event that the first self-reproducing cell contained a majority of L-amino acids over D-amino acids.

Another suggestion is that a comet 'seeded' the earth with L-amino acids and not D-amino acids, but this just 'pushes' the problem elsewhere.

The L-isomer of an amino acid is the naturally occurring one. Is there any obvious reason why it is naturally occurring rather than the D-isomer? If life started again or was found on a different planet, could it be right handed?

Answer

There is no obvious reason why the L-amino acid should be naturally occurring and not the D. It is quite likely that life could end up being 'right handed' if life started on a different planet.

Are there any advantages to living systems in only having one isomer? Consider how chiral molecules interact with enzymes (which are also chiral molecules).

Answer

If life was 'racemic' (i.e. chiral molecules existed as both enantiomers), then two sets of enzymes would be necessary in order to carry out the reactions of life. By concentrating on one enantiomer, life is more efficient.

In fact, the situation would be even more complex if life were racemic since there are an infinite number of ways in which L- and D- amino acids can combine to give an average sized enzyme containing a mixture of these enantiomers.

12.7 Chiral drugs

Many chiral drugs are natural products derived from sources such as plants and fungi. Indeed, the pharmaceutical industry started by extracting these natural products from such sources. As soon as the structures of these compounds were known, chemists tried to synthesise them in the laboratory. Chiral drugs such as morphine and quinine were successfully synthesised and had identical chemical, physical and spectroscopic properties to the naturally derived compounds.

However, there was one very important difference between the natural and the synthetic compounds. For example, the analgesic activity of synthetic morphine was half the analgesic activity of natural morphine, while the antimalarial activity of synthetic quinine was half the activity of natural quinine. The same story holds true for a large variety of natural chiral drugs which have been synthesised in the laboratory. Why should this be so?

Answer

Chiral drugs obtained from natural sources exist as a single enantiomer because their synthesis was catalysed by enzymes. Drugs synthesised in the laboratory were made using chemical reactions and were made as a racemic mixture of both enantiomers. A chiral drug such as morphine interacts with a chiral target in the body which can recognise one enantiomer of morphine but not the other. Therefore, only half the molecules in synthetic morphine would be pharmacologically active.

The pharmaceutical industry has recognised this problem for several years and is actively researching ways of synthesising the single enantiomer of chiral drugs—a process known as asymmetric synthesis.

12.8 (*R*) and (*S*) nomenclature

We have discussed enantiomers and described them as non-superimposable mirror images. We can identify which enantiomer is which by drawing their structures, but it would be better to define a particular enantiomer without needing to draw its structure. We can do this using the (*R*) and (*S*) nomenclature. The following example shows how this nomenclature works.

(Note that the process is quite similar to the process used to define alkenes as (*E*) or (*Z*), see Section 11.2.)

Example

Define the following lactic acid as (*R*) or (*S*).

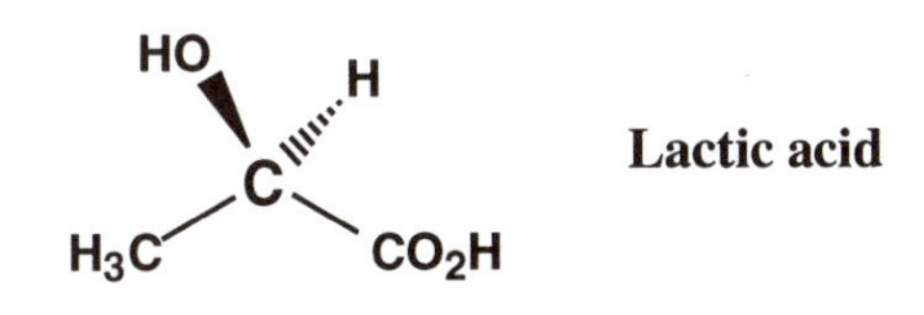

- Stage 1. Identify the asymmetric centre.

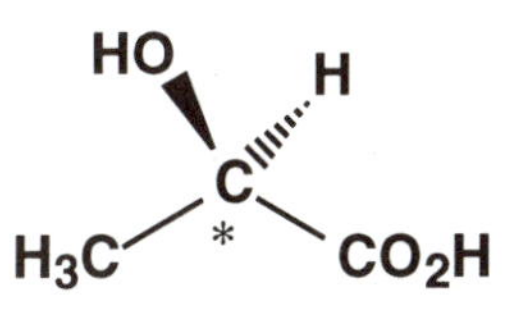

* Asymmetric centre

- Stage 2. Identify the atoms directly attached to the asymmetric centre and their atomic numbers.

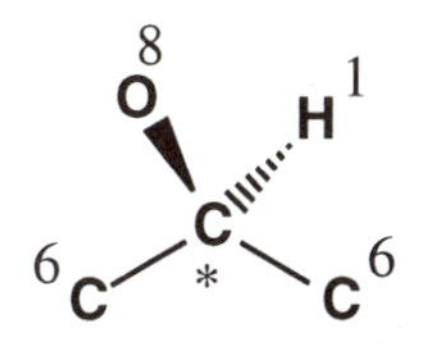

- Stage 3. Give the attached atoms a priority based on their atomic numbers.

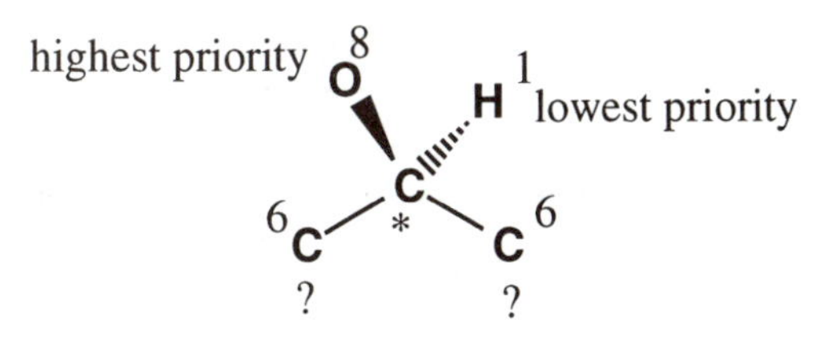

In this example, the two carbon atoms have the same atomic numbers and cannot be given a priority.

- Stage 4. Choose the priority of any identical atoms by moving to the next atom of highest atomic number.

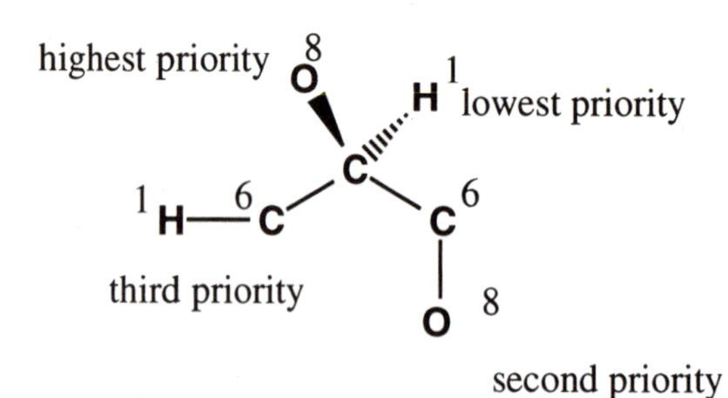

- Stage 5. Place the group of lowest priority 'behind the page' and draw an arc connecting the remaining groups, starting from the group of highest priority and finishing at the group of third priority.

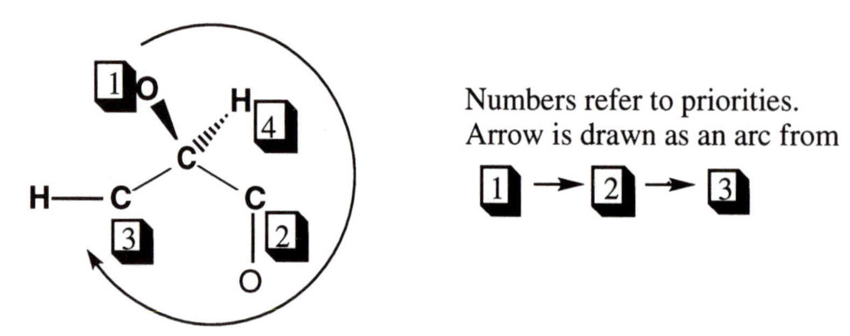

In this example, the hydrogen is the group of lowest priority and is already 'behind the page'.

- Stage 6. If the arc is drawn clockwise, the assignment is (*R*). If the arrow is drawn anticlockwise, the assigment is (*S*).
  In this example the arrow is drawn clockwise. Therefore, the molecule is (*R*)-lactic acid.

Assign the following amino acid as (*R*) or (*S*).

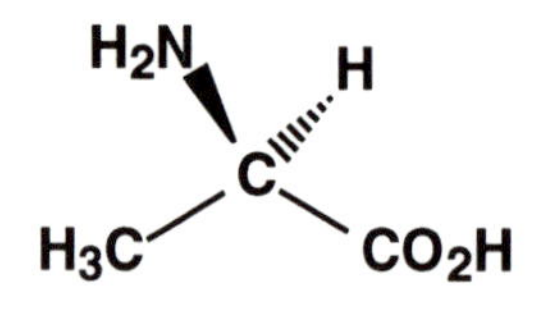

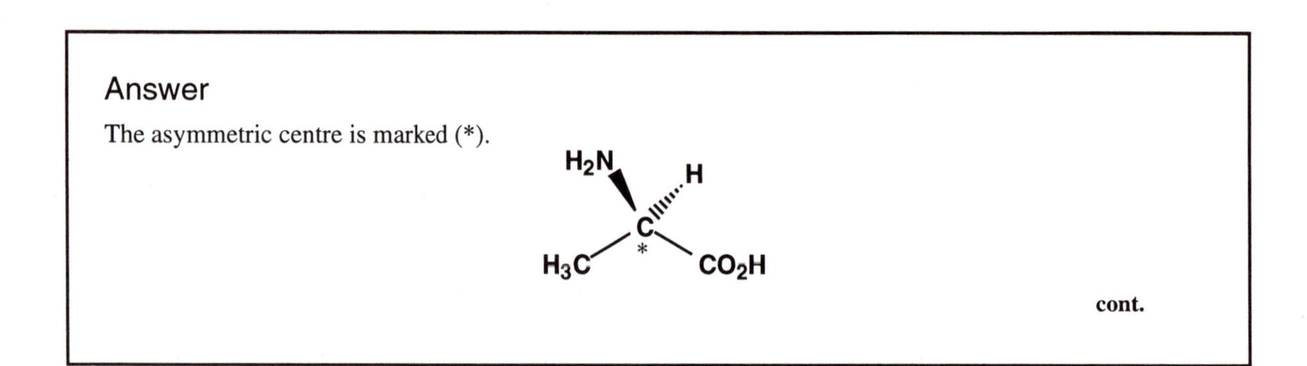

Answer

The asymmetric centre is marked (*).

cont.

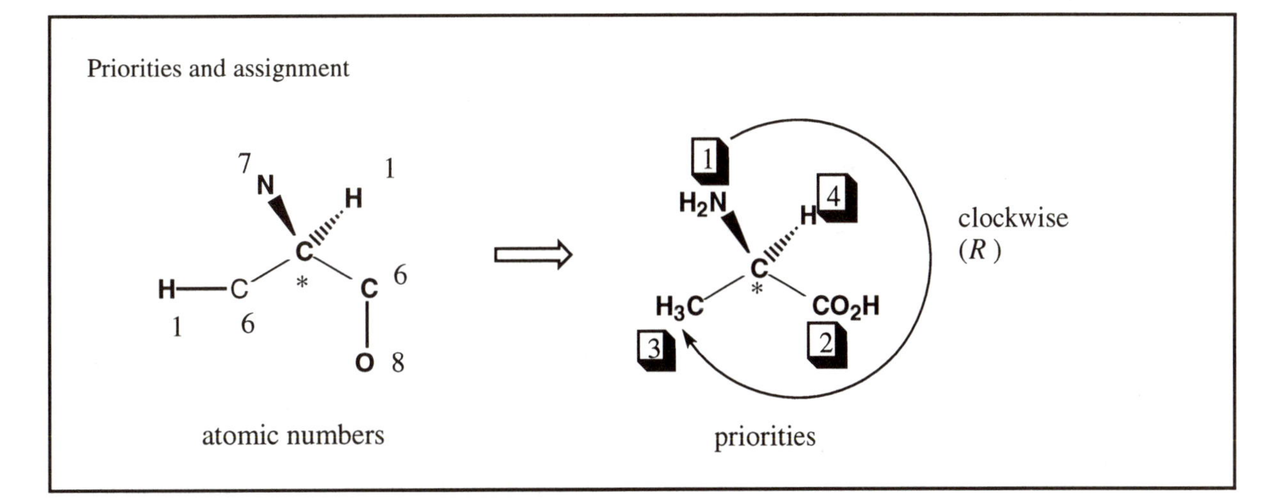

This next structure is a bit trickier to assign. Identify the asymmetric centre and the atoms directly attached to it. Give the priorities which are possible at this stage.

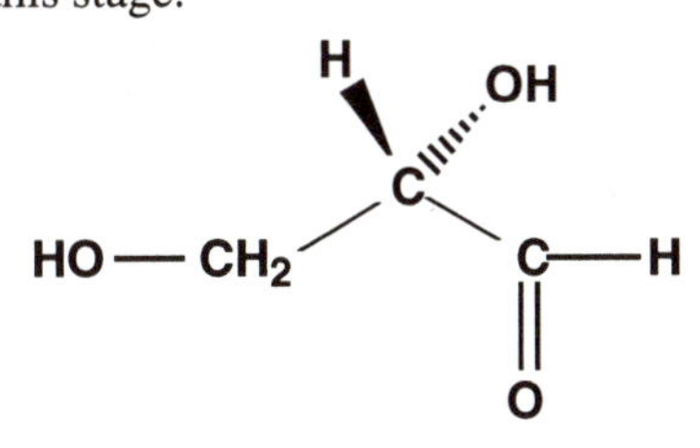

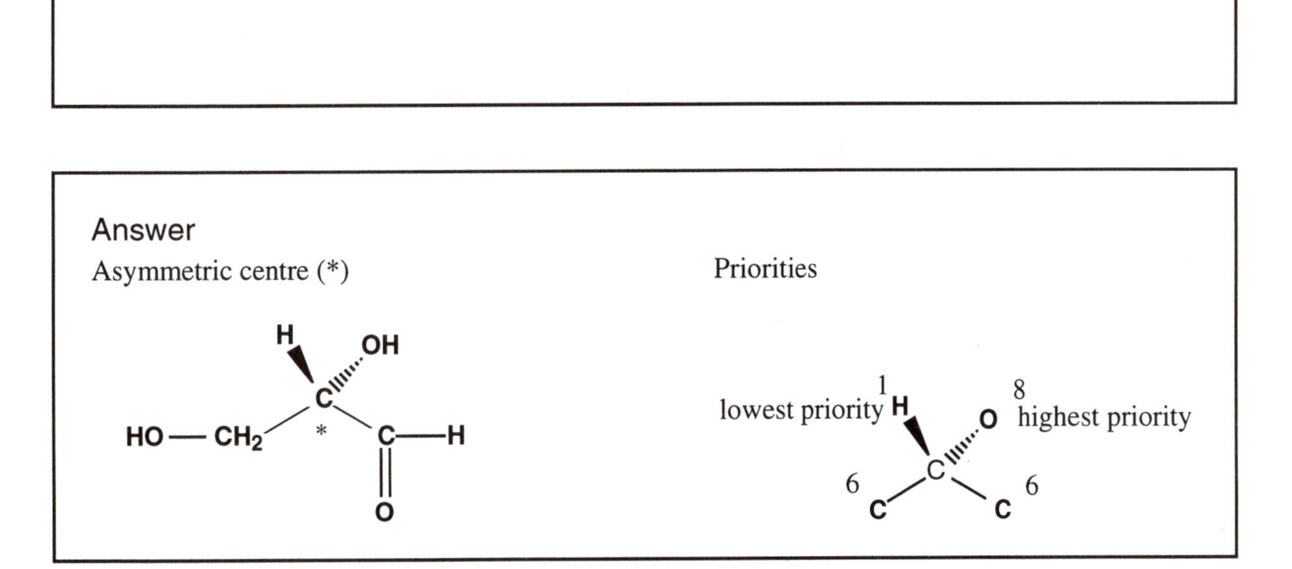

We have to move to the next stage if we want to distinguish between two identical atoms. Carry out the next stage in the procedure and see if the relevant groups can be given a priority.

Answer

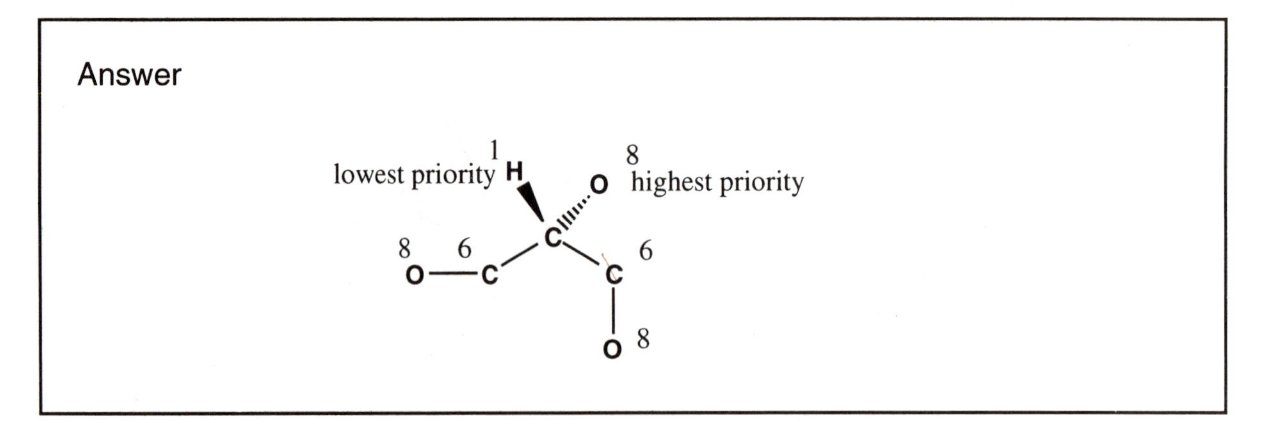

This next stage still does not distinguish between the CHO and CH_2OH groups since both carbon atoms have an oxygen atom attached. In situations like this, you now look at the second most important atom attached to the two carbon atoms. If there is a double bond present, you are allowed to 'visit' the same atom twice.

Apply these rules to the example and see if you can give a priority to the two groups.

Answer

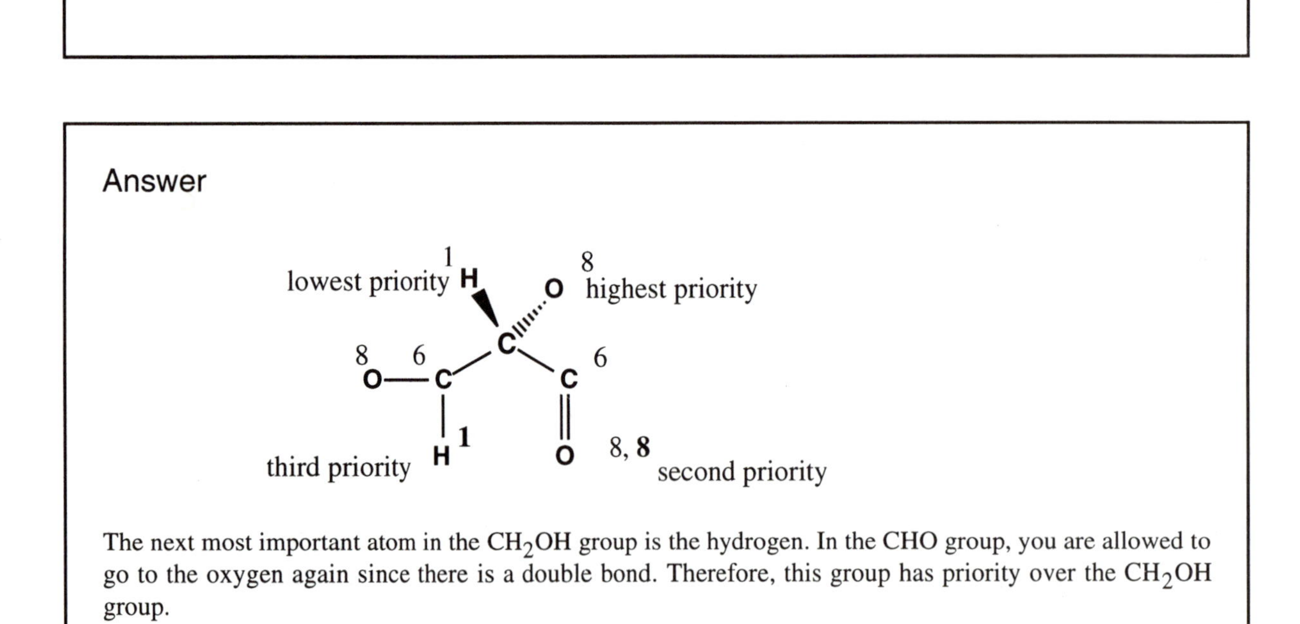

The next most important atom in the CH_2OH group is the hydrogen. In the CHO group, you are allowed to go to the oxygen again since there is a double bond. Therefore, this group has priority over the CH_2OH group.

Assign the molecule as (*R*) or (*S*).

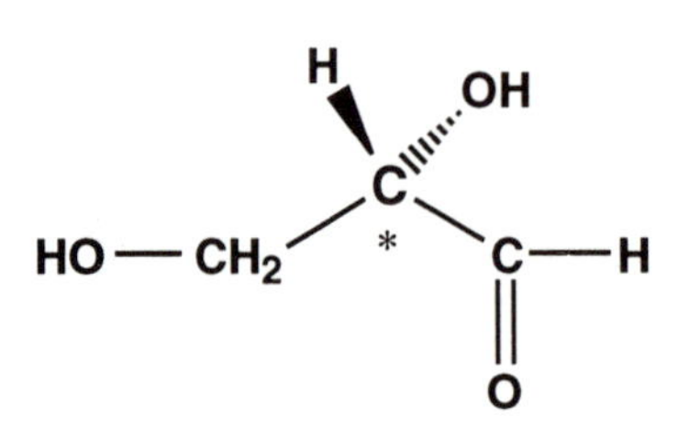

Answer

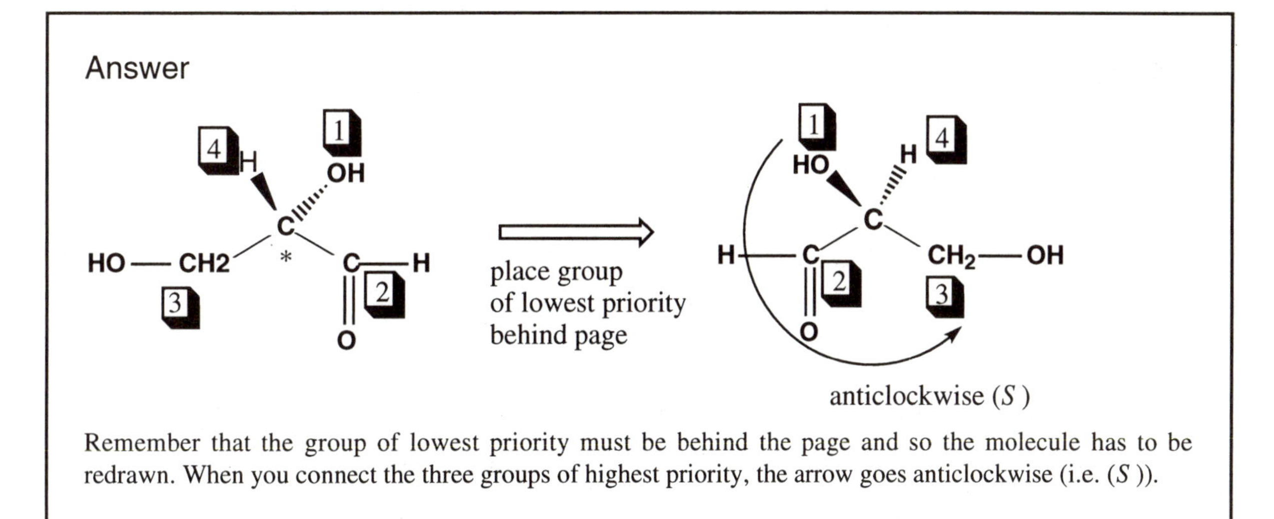

Remember that the group of lowest priority must be behind the page and so the molecule has to be redrawn. When you connect the three groups of highest priority, the arrow goes anticlockwise (i.e. (*S*)).

12.9 Molecules to beware of!

In Section 12.2, we stated that it was possible for:

(a) some chiral molecules to have no asymmetric centres;

(b) some molecules to have more than one asymmetric centre and yet not be chiral.

The following two molecules are examples of molecules which have no asymmetric centre but which are still chiral.

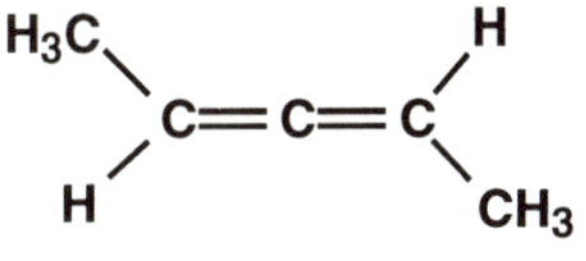

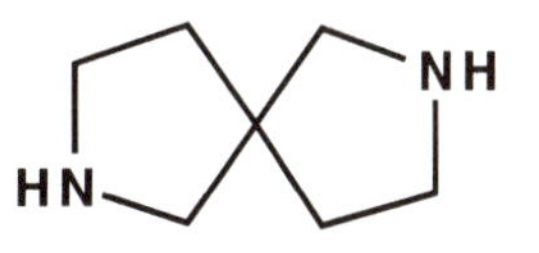

An allene

A spiro structure

It is not easy to see why these structures are chiral just by looking at them on the page, and you will have to construct models to prove to yourself that they are chiral. Construct models for both the allene and the spiro compound. Study the models of these molecules and decide whether the drawings above are accurate representations of their shapes. If they are not, draw better representations for the allene and the spiro compound.

Answer

The structures drawn previously are not good representations of the allene or the spiro compound since they imply that the molecules are planar. Models show that the substituents at either end of the allene are in different planes, and that the rings in the spiro structure are at right angles to each other.

H_3C, H — C=C=C — H, CH_3

An allene

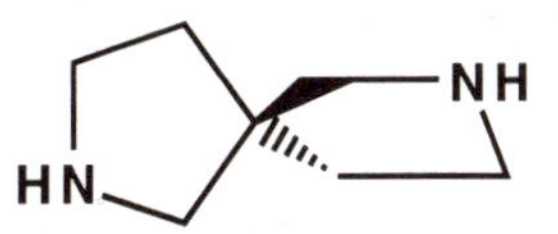

A spiro structure

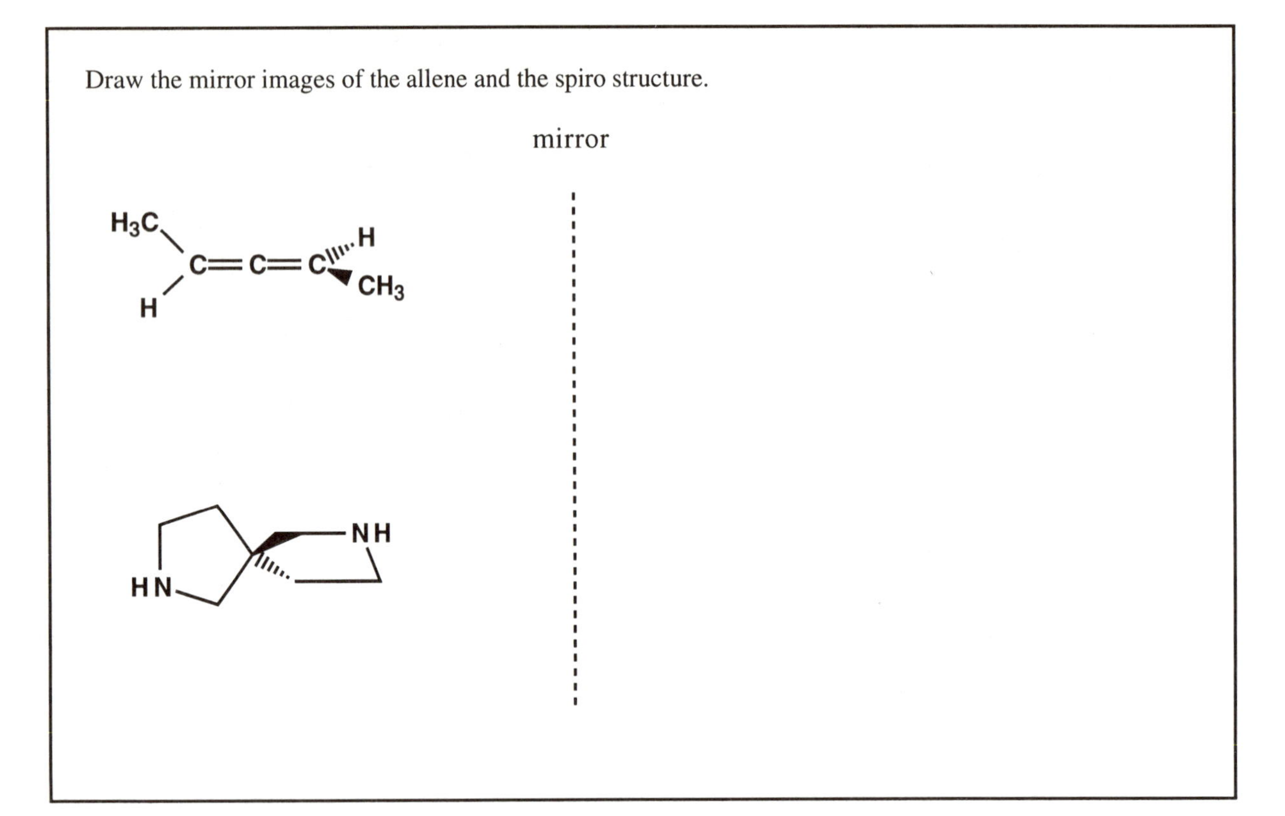

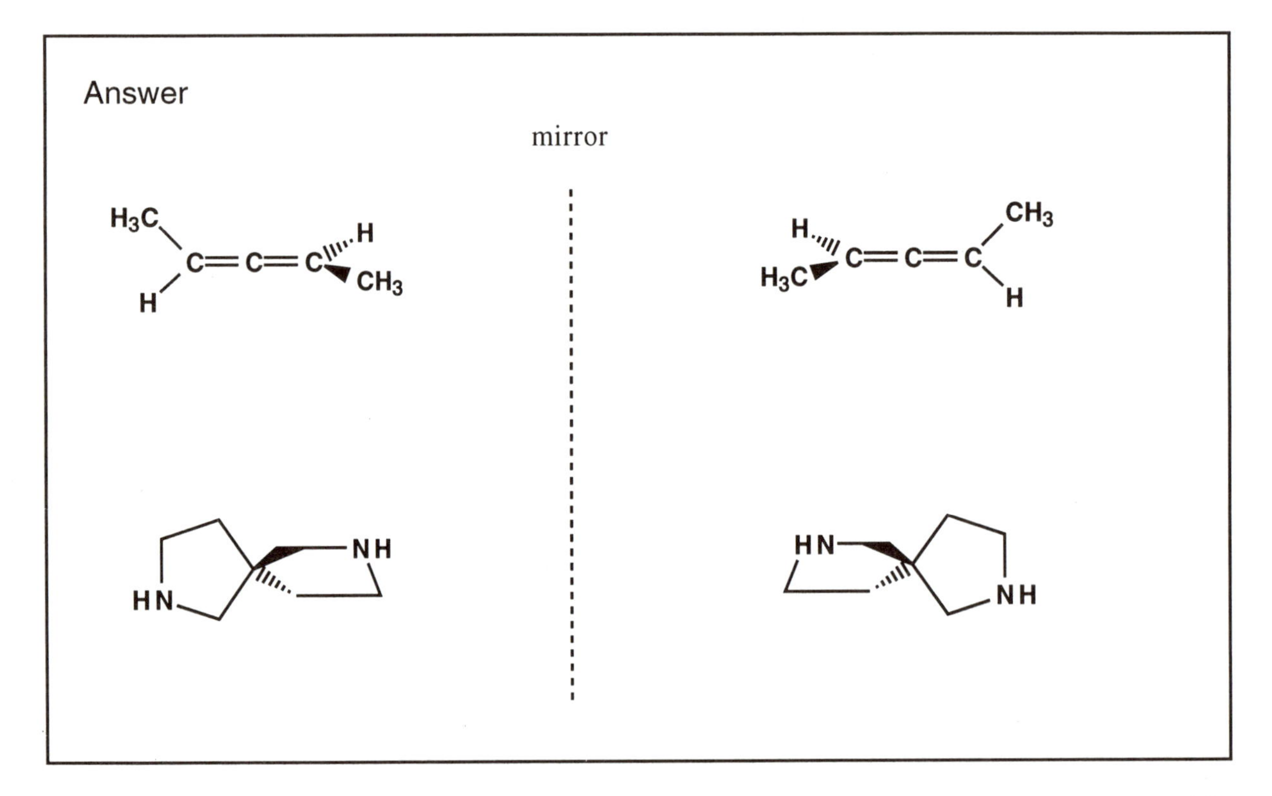

Construct both mirror images of the allene and both mirror images of the spiro structure. Are the models superimposable?

Answer

They are not superimposable and therefore the allene and the spiro structures are chiral compounds which can exist as two non-superimposable mirror images or enantiomers.

These examples demonstrate that the presence of an asymmetric centre is not required for a molecule to be chiral.

A better rule for determining whether a molecule is chiral or not is to study the symmetry of the molecule. A molecule will be chiral if it is asymmetric (i.e. has no elements of symmetry). It is also possible for a molecule to be chiral if it has an axis of symmetry, but no other elements of symmetry. The allene already mentioned is an example of this, since it contains an axis of symmetry through its central carbon.

We have seen two molecules which are chiral and yet have no asymmetric centres. The following molecule has two asymmetric centres but is not chiral. Identify the two asymmetric centres.

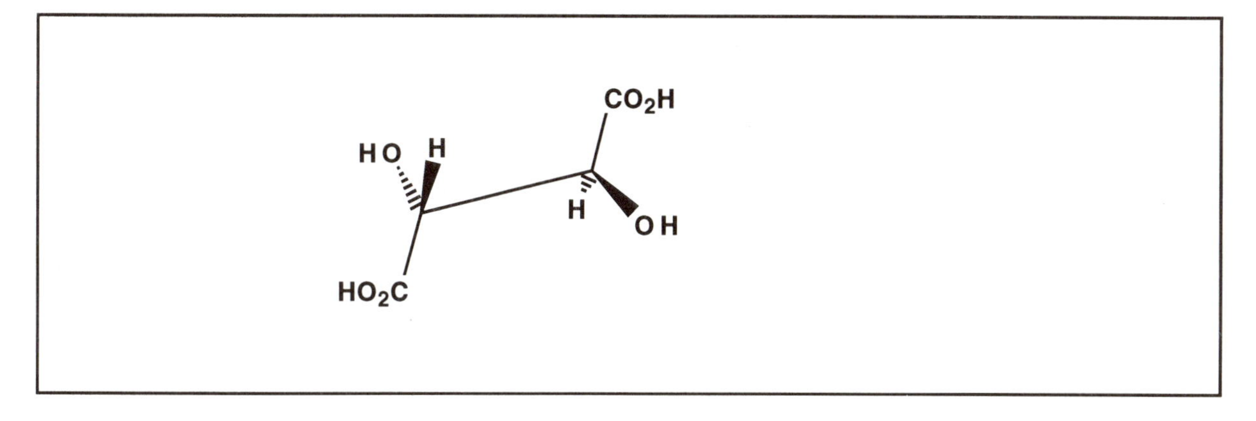

Answer

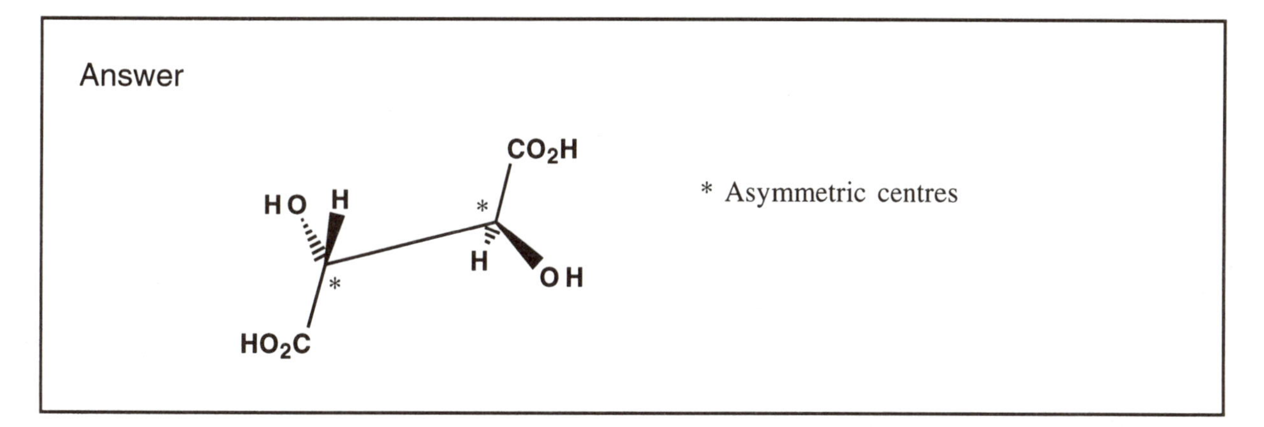

You can prove to yourself that this molecule is not chiral by constructing models of both mirror images and showing that the models are superimposable.

Since the molecule is not chiral, it must contain some element of symmetry. It has in fact a plane of symmetry. This is not obvious in the structure as it is drawn above, but the molecule can be rotated around the central C-C bond to give a variety of shapes (known as conformations). Study a model of the above structure and rotate the molecule round the central bond until you find a shape or conformation which results in a plane of symmetry, then draw the conformation below.

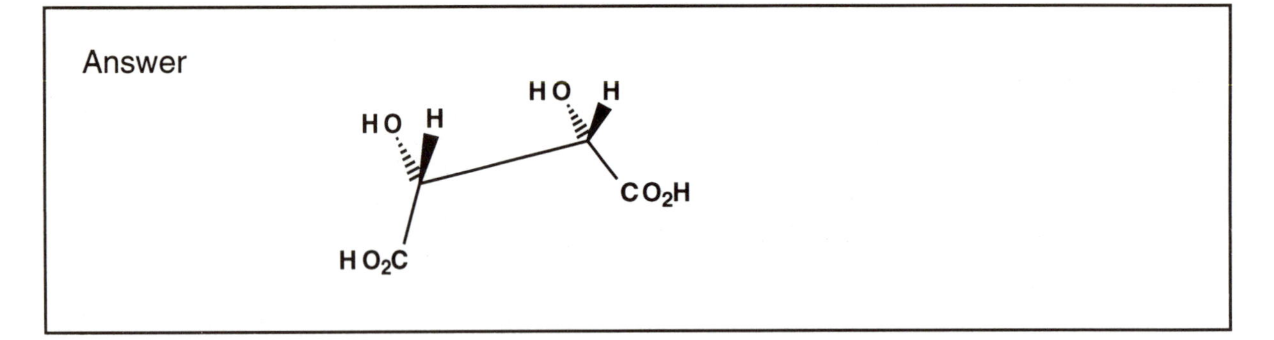

Draw the plane of symmetry for this conformation.

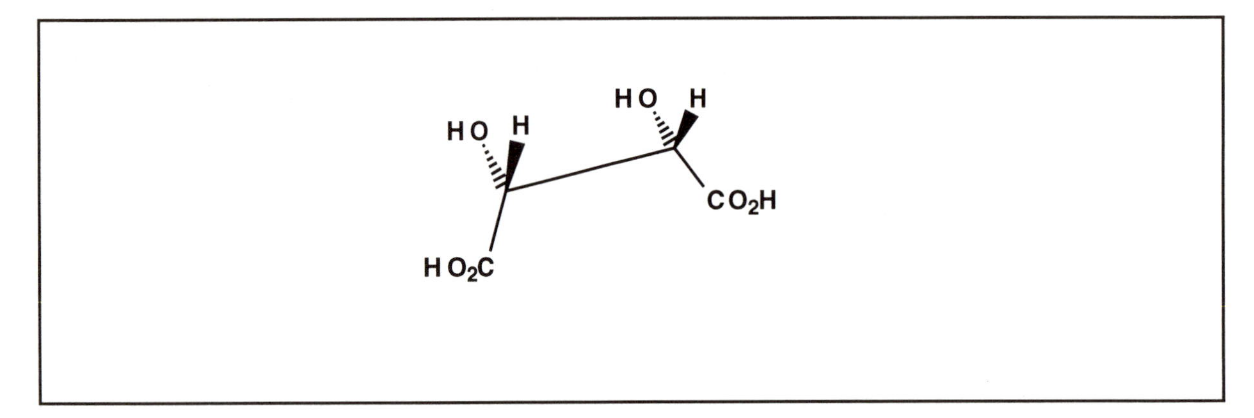

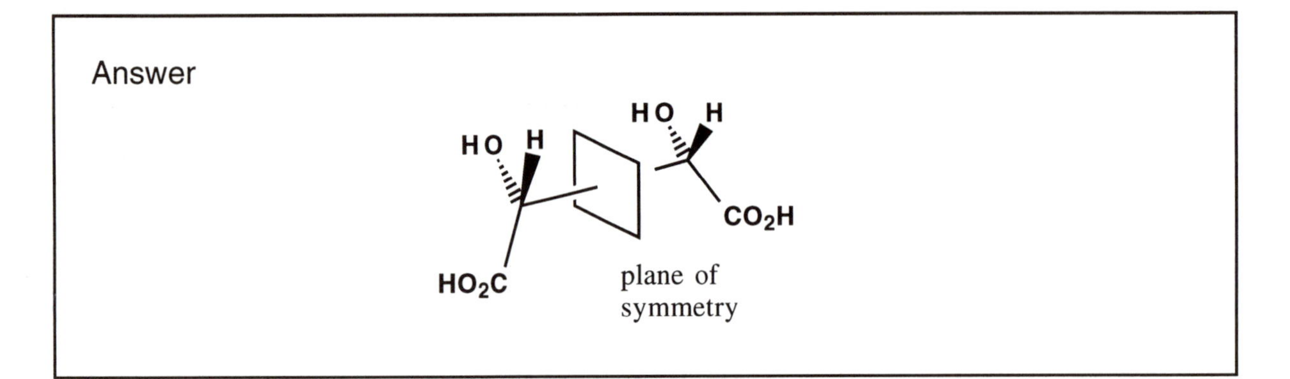

Summary

- Asymmetric or chiral molecules can exist as two non-superimposable molecules known as enantiomers.
- Enantiomers have identical chemical and physical properties with two exceptions:

    (a) the direction in which they rotate plane polarised light,

    (b) their interaction with other chiral molecules such as enzymes.
- Enantiomers are configurational isomers.
- A mixture of enantiomers is known as a racemate.
- The chiral molecules present in the human body exist as one enantiomer.
- An asymmetric centre is a centre with four different substituents.
- The configuration of an asymmetric centre can be defined as (*R*) or (*S*).
- Not all chiral molecules have asymmetric centres.
- Not all molecules having asymmetric centres are chiral.

SECTION 13

Conformational Isomers

Conformational Isomers

We have looked at constitutional and configurational isomerism. There is a third type of isomerism called conformational isomerism. Conformational isomers are essentially different shapes of the same molecule resulting from rotation round single bonds. Since rotation round a single bond normally occurs easily at room temperature, conformational isomers are not different compounds and are freely interconvertable. They cannot be separated from each other in the way constitutional and configurational isomers can.

13.1 Conformations of alkanes

It is a good idea to use models for this section. Look at the structure of ethane and draw two different shapes for this molecule.

Answer

There are many different shapes which could be adopted by ethane, but the most distinctive ones are the ones shown below.

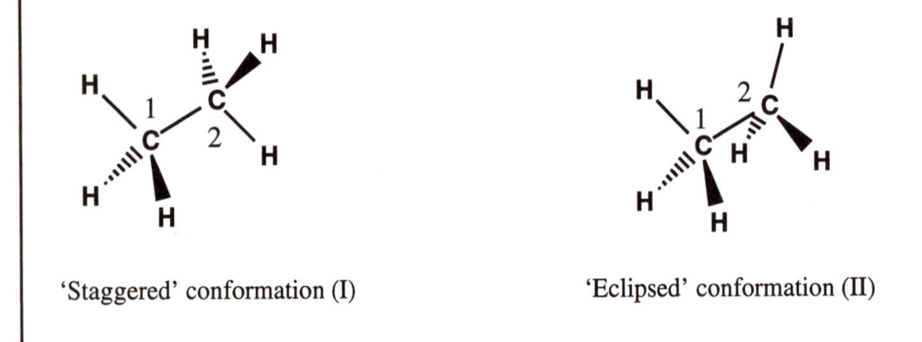

'Staggered' conformation (I)

'Eclipsed' conformation (II)

The two conformations I and II above are called 'staggered' and 'eclipsed' respectively. Why?

Answer

If you compare the C–H bonds on carbon 1 with the C–H bonds on carbon 2, then they are staggered with respect to each other in conformation I and eclipsed in conformation II.

These conformations can be drawn in a different way to emphasise this difference. If you imagine looking at conformation II along the C1–C2 bond, you could draw the structure as follows.

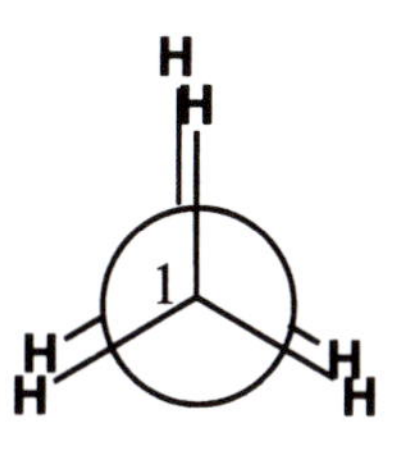

This is known as a Newman projection. Carbon 2 is hidden behind carbon 1 and the C–H bonds are eclipsed. Now draw the Newman projection for the staggered conformation of ethane.

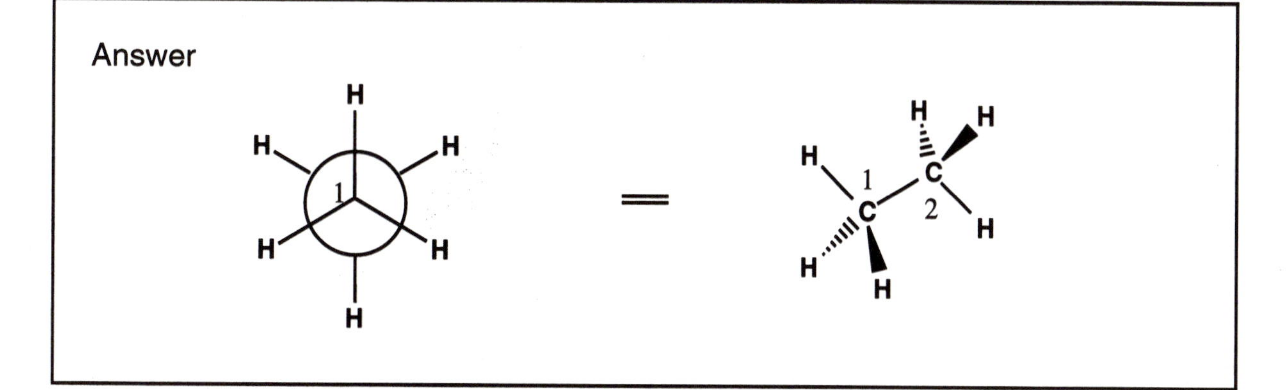

Which of these conformations do you think would be the most stable and why?

Answer

The staggered conformation is the more stable conformation. In this conformation, the C–H bonds and the hydrogen atoms themselves are well separated. In the eclipsed conformation, they are closer together and this causes strain.

Eclipsed conformations are not favoured because of the unfavourable strains resulting from eclipsed bonds and atoms. Therefore, the most likely conformations in a molecule are the 'staggered' conformations. Ethane has only one such conformation, but different staggered conformations are possible with larger molecules.

Make a molecular model of butane, then identify and draw the two possible staggered conformations for this molecule.

Answer

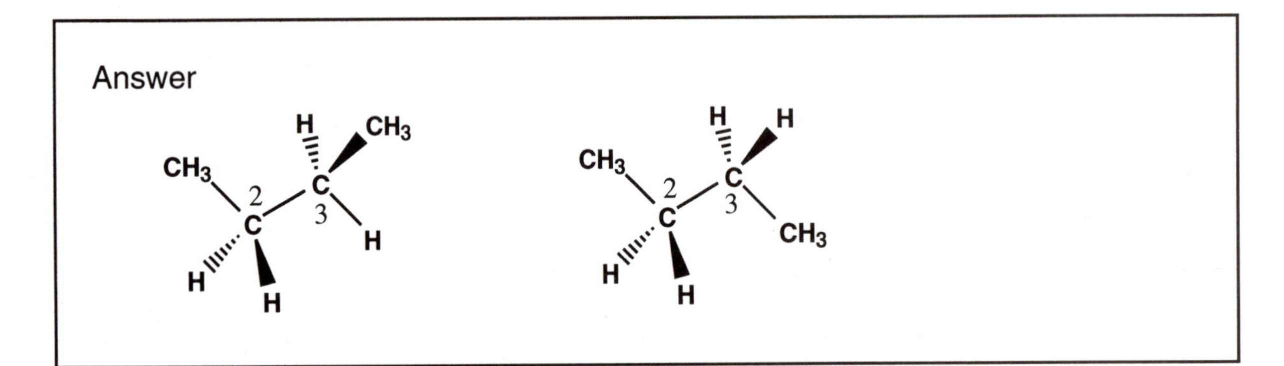

Draw the Newman projections for these structures by looking along the C2–C3 bond.

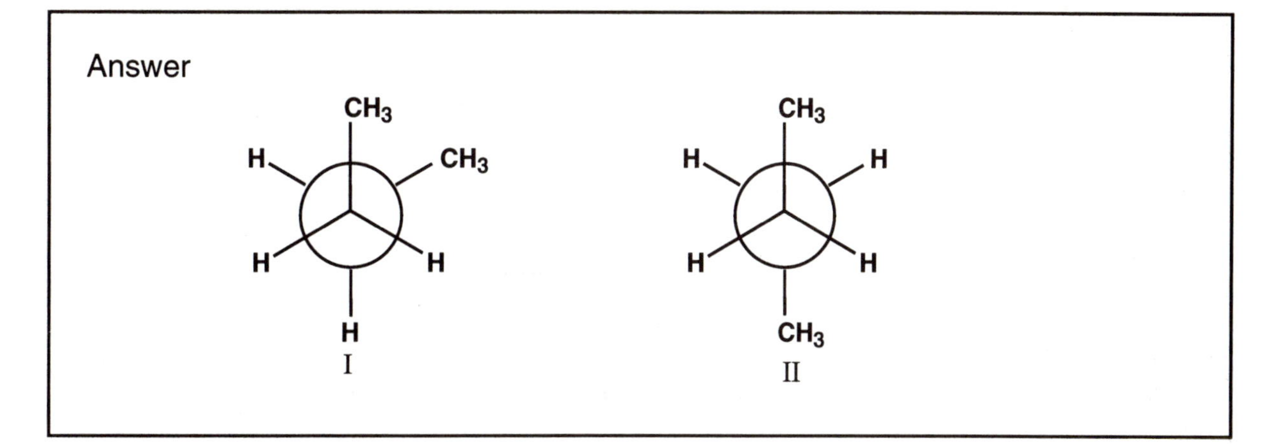

One of these isomers is more stable than the other. Which do you think is the more stable and why?

Answer

Isomer II is the more stable. This is because the methyl groups in this conformer are as far apart from each other as possible. The methyl groups are bulky and in conformation I are close enough to interact with each other and lead to some strain.

If you consider the carbon atoms only, the most stable conformation for butane is a 'zigzag' shape.

In this shape, the carbon atoms are as far apart from each other as possible.
Taking this into account, draw the most stable conformations for hexane, octane and decane.

Answer

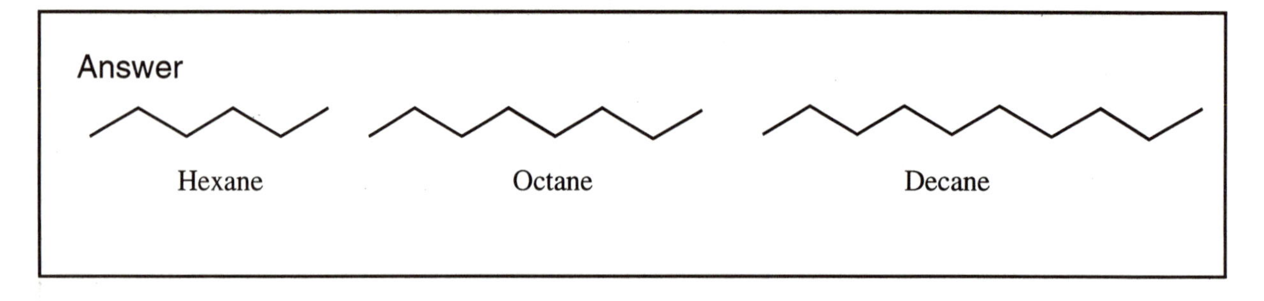

13.2 Conformations of cyclic structures

Build molecular models of benzene and cyclohexane and compare their shapes. Is there any difference?

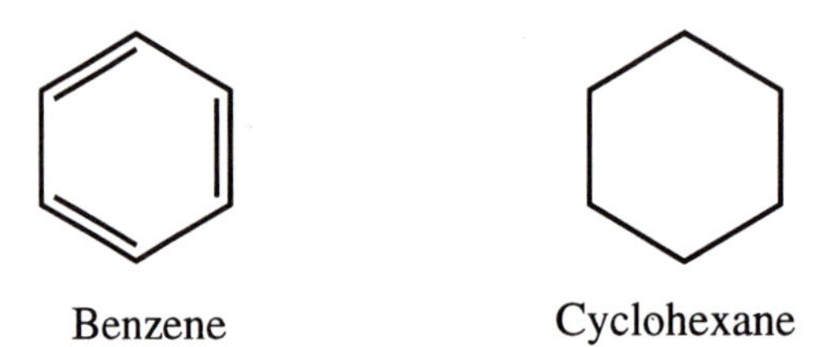

Answer

There is a marked difference in shape between these two molecules. Benzene is completely planar whereas cyclohexane is 'puckered'.

The above exercise emphasises the importance of making models to find out the actual shape of a molecule. When drawn on the page, both benzene and cyclohexane look flat. As you have seen, there is very little similarity in shape between the two molecules.

Construct molecular models of cyclopropane, cyclobutane and cyclopentane and try to draw a more accurate diagram to demonstrate their shape or shapes.

Cyclopropane

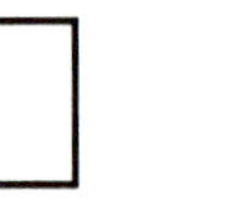

Cyclobutane

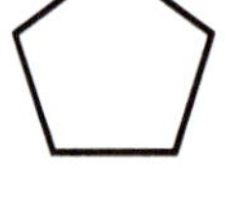

Cyclopentane

Cyclopropane

cont.

Cyclobutane

Cyclopentane

Answer

Cyclopropane is a flat molecule with respect to the carbon atoms. The hydrogen atoms are above and below the plane of the ring.

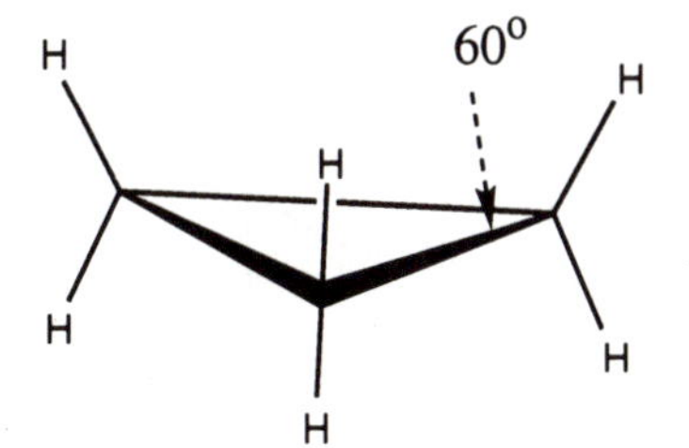

Cyclobutane can form three distinct shapes—a planar shape and two 'butterfly' shapes.

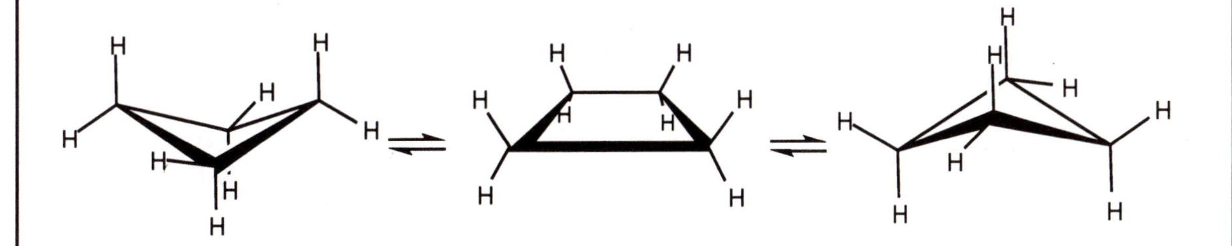

Cyclopentane can also form a variety of shapes or conformations.

cont.

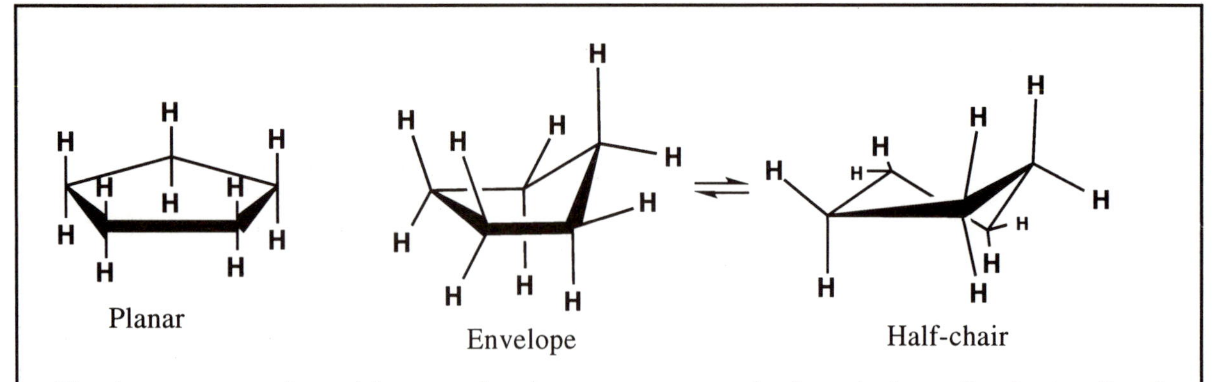

The planar structures for cyclobutane and cyclopentane are too strained to exist in practice due to eclipsed C–H bonds. You can check this by studying models of these rings.

The two main conformational shapes for cyclohexane are a chair shape and a boat shape.

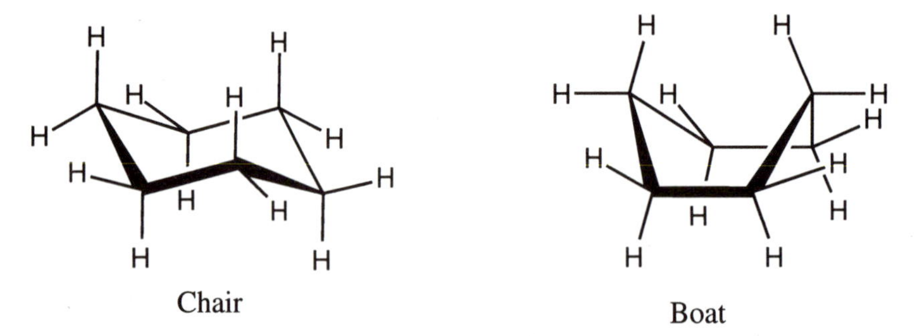

The chair is more stable than the boat since the latter has eclipsed C–C and C–H bonds.

This can be seen better from the Newman projections of both conformations. These have been drawn below for both the chair and the boat conformations and have been drawn such that you are looking along two bonds at the same time. The carbon atoms for the chair and the boat conformations have been numbered and carbon number two has been identified in the Newman projections. Identify the remaining carbon atoms in the Newman projections and identify which bonds are being 'looked along'.

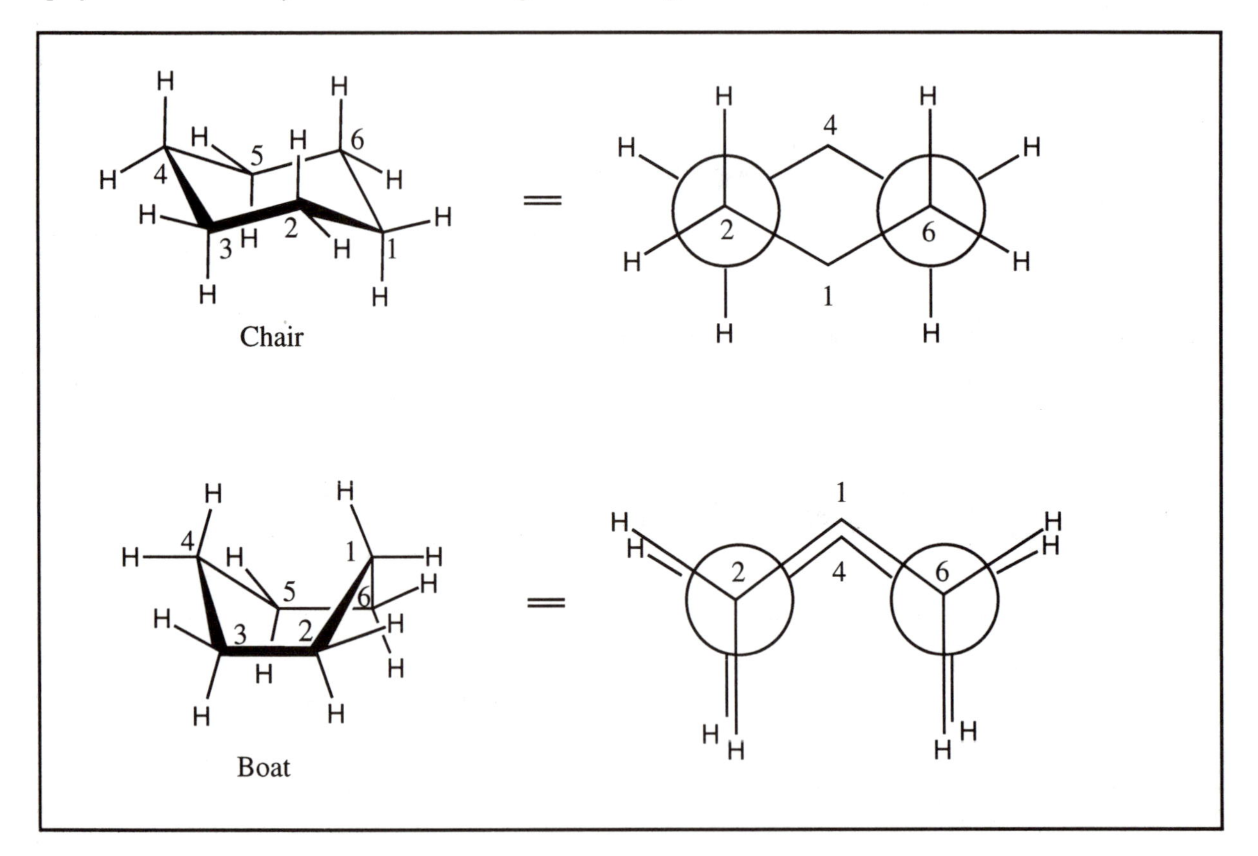

Answer

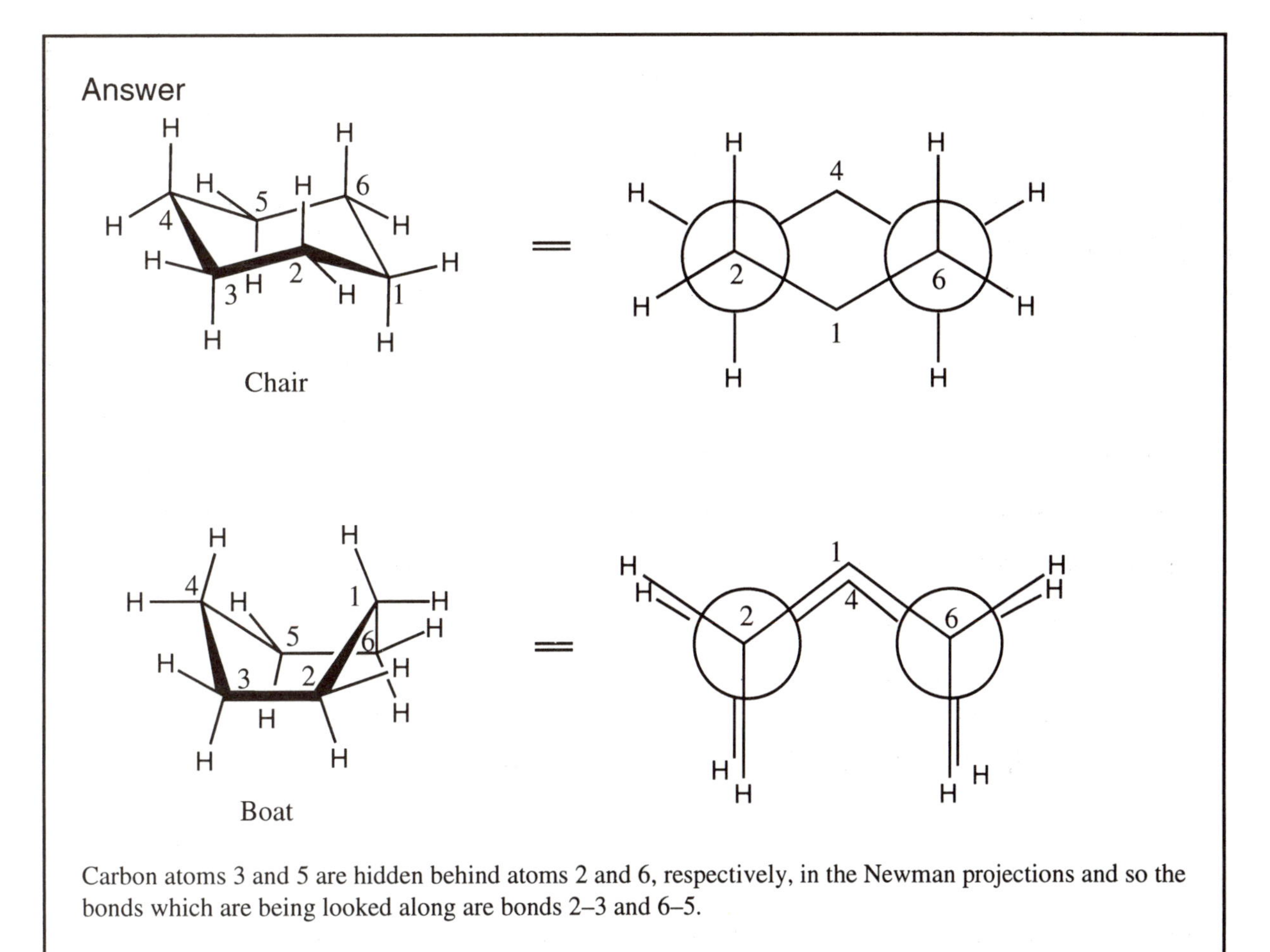

Carbon atoms 3 and 5 are hidden behind atoms 2 and 6, respectively, in the Newman projections and so the bonds which are being looked along are bonds 2–3 and 6–5.

Identify the C–C bonds which are eclipsed in the boat conformation.

Answer

Bond 1–2 is eclipsed with bond 3–4. Bond 1–6 is eclipsed with bond 5–4.

Summary

- Conformational isomers are different shapes of the same molecule resulting from rotation around a single bond.
- The most stable conformation for a particular molecule is the conformation with the least steric and electronic strain.
- The preferred conformation for straight-chain alkanes is a zigzag shape.

SECTION 14

Shape and Bonding

Shape and Bonding

14.1 Atomic structure of carbon

We have looked at the shape of a range of organic molecules and we have also seen how different functional groups have different shapes. We now have to explain why. Why do alkanes have tetrahedral carbons? Why are aromatic rings flat? Why are alkenes and carbonyl groups planar and why are alkyne groups linear? In order to understand the shape of organic molecules we have to understand bonding, and in order to understand that we have to look first at the atomic structure of carbon and in particular its electronic structure.

Take a look at the periodic table. How many electrons does carbon have?

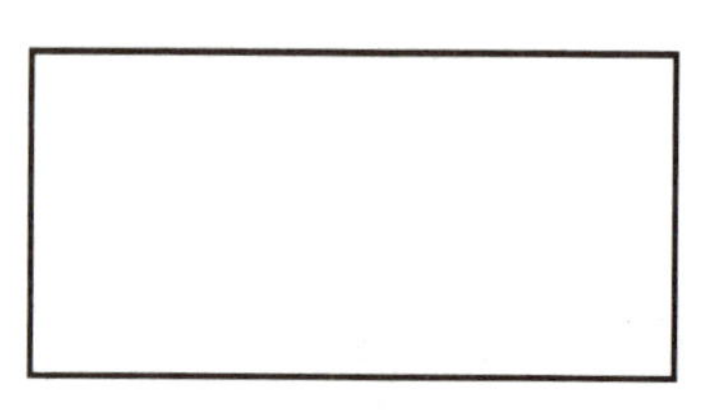

Answer

Carbon has six electrons.

Are all these electrons of equal energy? If not, why not?

Answer

The six electrons are not of equal energy since they occupy different atomic orbitals of different energies.

Carbon is in row 2 of the periodic table. This means that there are two shells of atomic orbitals. The first shell is closest to the nucleus and only has a single s orbital — the 1s orbital. The second shell has a single s orbital (2s) and three p orbitals (3 × 2p).

The s orbitals are spherical in shape with the 2s orbital much larger then the 1s orbital. The p orbitals are dumb-bell shaped and are aligned along the three different axes such that there are $2p_x$, $2p_y$ and $2p_z$ atomic orbitals.

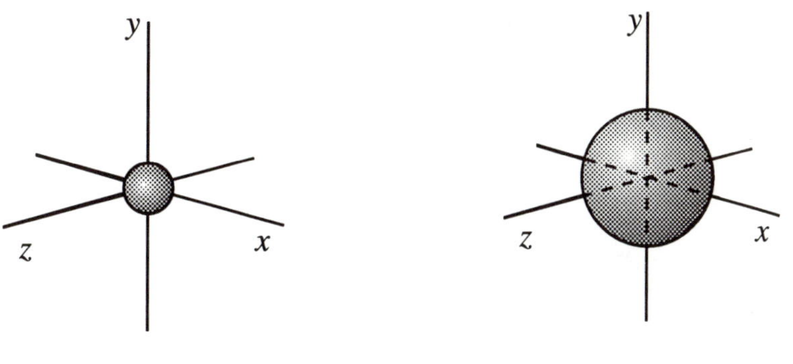

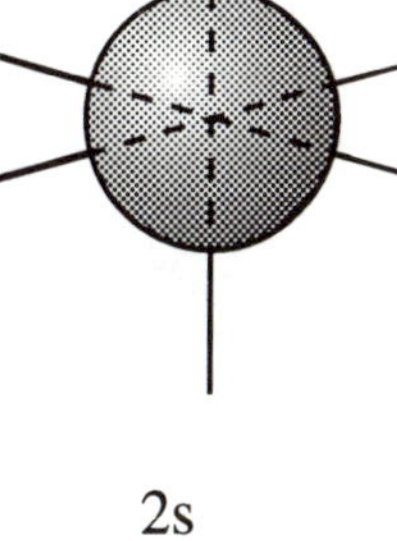

1s

2s

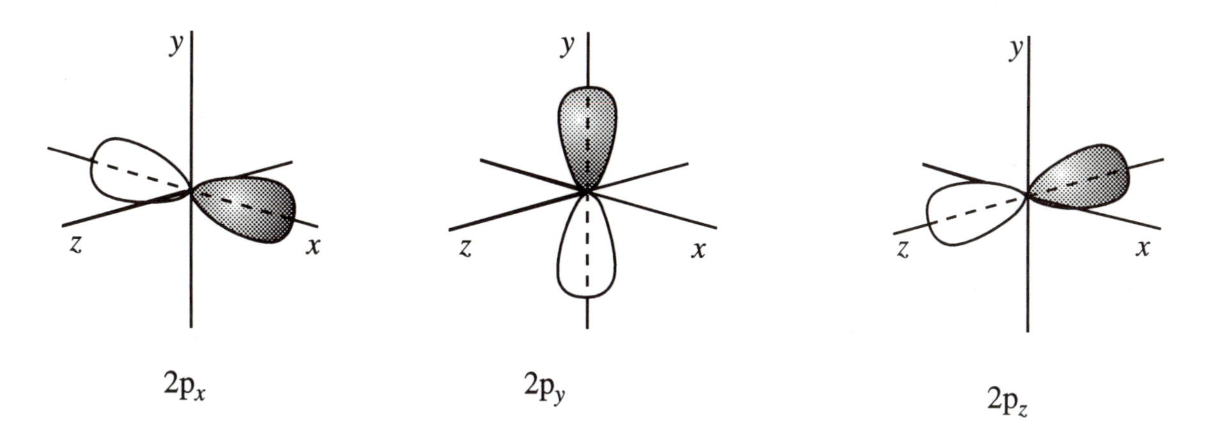

$2p_x$ $2p_y$ $2p_z$

These orbitals are not of equal energy. The 1s orbital has the lowest energy. The 2s orbital has a higher energy and the 2p orbitals have the highest energies.

Draw a diagram to represent the different energy levels of these orbitals.

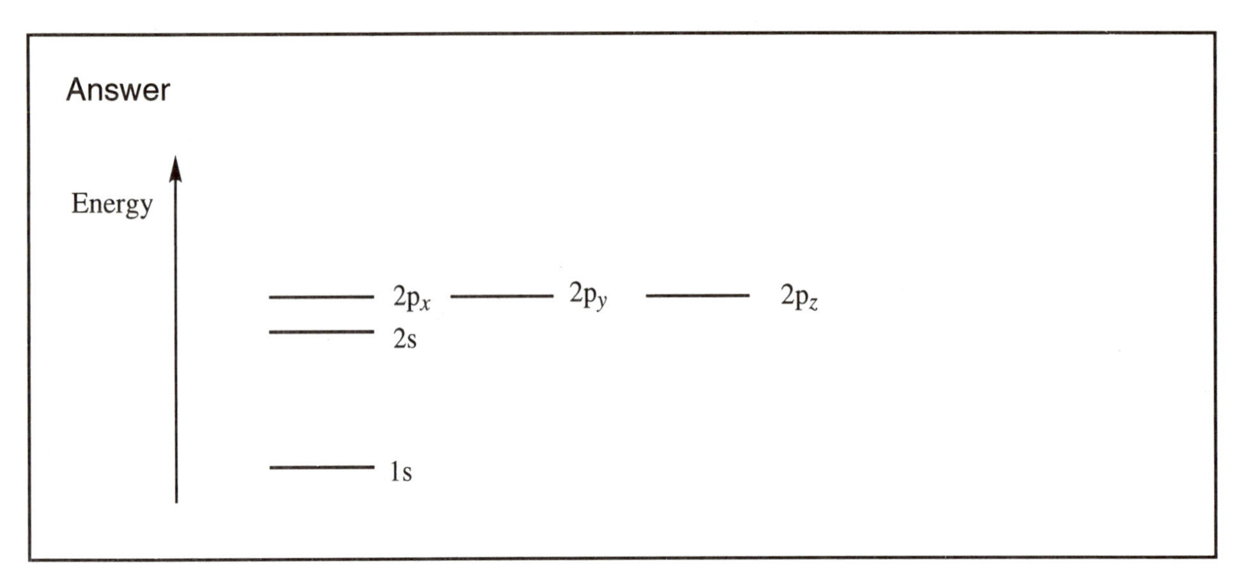

Since the atomic orbitals have different energies, in what order are they filled up by electrons?

Answer

The atomic orbitals of lowest energy are filled up first. Therefore, the 1s orbital will be filled up before the 2s orbital, which will be filled up before the 2p orbitals.

We are now ready to fit in four of carbon's six electrons. Redraw the energy levels and fit the first four electrons into their atomic orbitals. Remember that each orbital is allowed a maximum of two electrons and the electrons in a filled orbital have opposite spins.

Answer

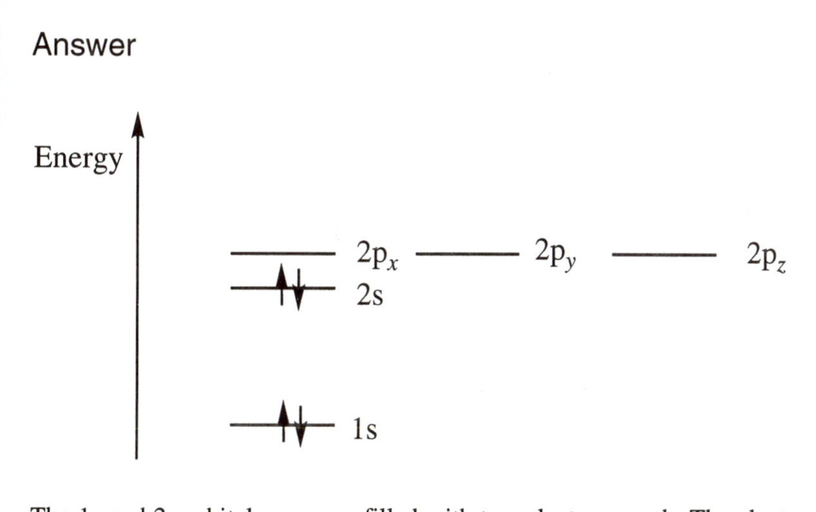

The 1s and 2s orbitals are now filled with two electrons each. The electrons in each orbital have opposite spins and this is represented by the arrows pointing up or down.

You have two electrons left to fit into the 2p orbitals. You now have a decision to make. You have three p orbitals of equal energy. You can either place the two electrons into a single p orbital to make a full p orbital or you can place them into two separate p orbitals to make two half-filled p orbitals. Which is correct?

Answer

The two electrons go into two separate p orbitals to make two half-filled p orbitals. This is a general rule. Whenever you have orbitals of equal energy, each electron will go into a different orbital before pairing begins. This is known as Hund's rule.

Draw the full energy level diagram for the carbon atom.

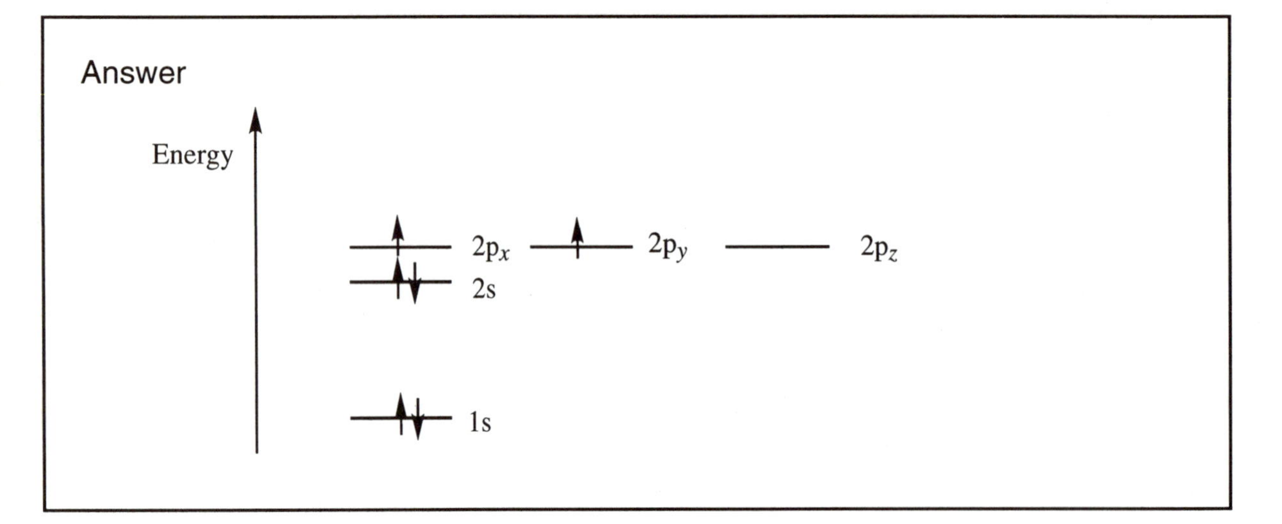

The electronic configuration for the carbon atom can be described as $1s^22s^22p_x{}^12p_y{}^1$. The numbers in superscript refer to the numbers of electrons in each orbital. The letters refer to the types of atomic orbital and the numbers in front refer to which shell the orbitals belong.

14.2 Bonding

A bond is what binds two atoms together within a structure. It is formed by the sharing of a pair of electrons between these atoms. Atoms can form bonds by sharing unpaired electrons.

How many unpaired electrons does a carbon atom have and how many bonds would you expect carbon to make with other carbon atoms?

Answer

There are two unpaired electrons and therefore you would expect carbon to form two bonds.

14.3 Hybridisation

Hopefully, you will now be puzzled since you should know that carbon always forms four bonds. It is clear that the simple atomic structure picture of carbon does not fit the facts. How then does a carbon atom form four bonds with only two unpaired electrons?

The first thing to appreciate is that the picture we have drawn of carbon's atomic structure is of an isolated carbon atom which is not bonded to anything. When a carbon atom is part of a molecular structure, the picture changes and we do not have the same orbitals as in the isolated atom. When carbon is forming bonds, it can 'mix' the s and p orbitals in the second shell (valence shell). This is known as hybridisation and it allows carbon to form the four bonds which we observe in reality.

There are three ways in which an s orbital can be mixed with some or all of the p orbitals. Can you identify what the three forms of hybridisation could be?

Answer

- The 2s orbital is mixed with all three 2p orbitals. This is known as sp^3 hybridisation.
- The 2s orbital is mixed with two of the 2p orbitals. This is known as sp^2 hybridisation.
- The 2s orbital is mixed with one of the 2p orbitals. This is known as sp hybridisation.

We shall look at each of these types of hybridisation in turn.

14.4 sp^3 hybridisation

In sp^3 hybridisation, the 2s orbital is mixed with all three of the 2p orbitals to give a set of hybrid orbitals. The number of hybrid orbitals must equal the number of original atomic orbitals used for mixing. The hybrid orbitals will each have the same energy but be different in energy from the original atomic orbitals. That energy difference will reflect the mixing of the respective atomic orbitals.

Predict the number of sp^3 hybridised orbitals and draw an energy diagram comparing the energies of the original atomic orbitals and the resulting sp^3 hybridised orbitals.

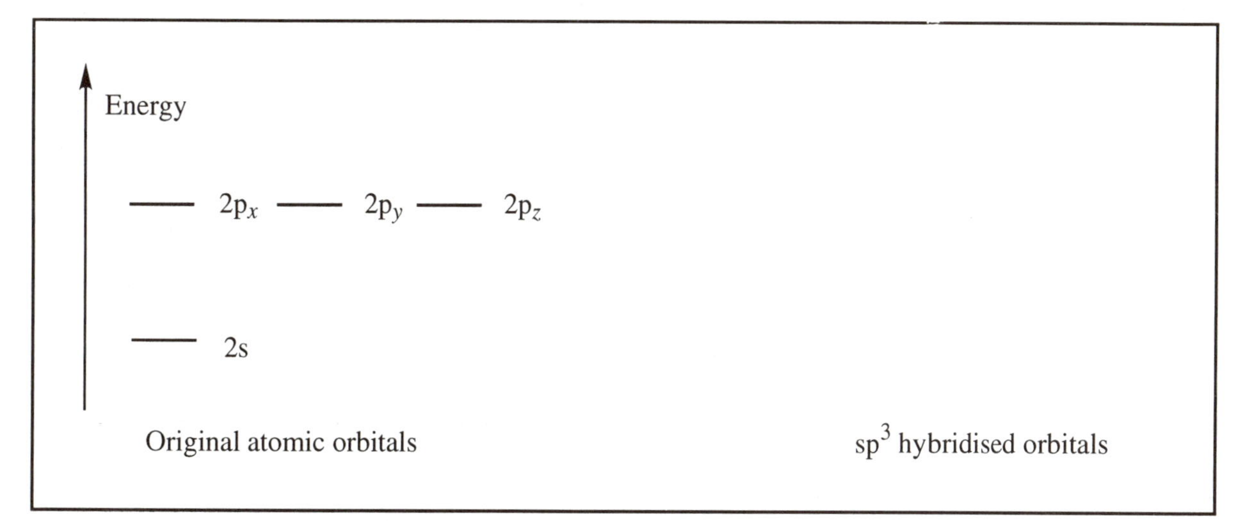

Answer

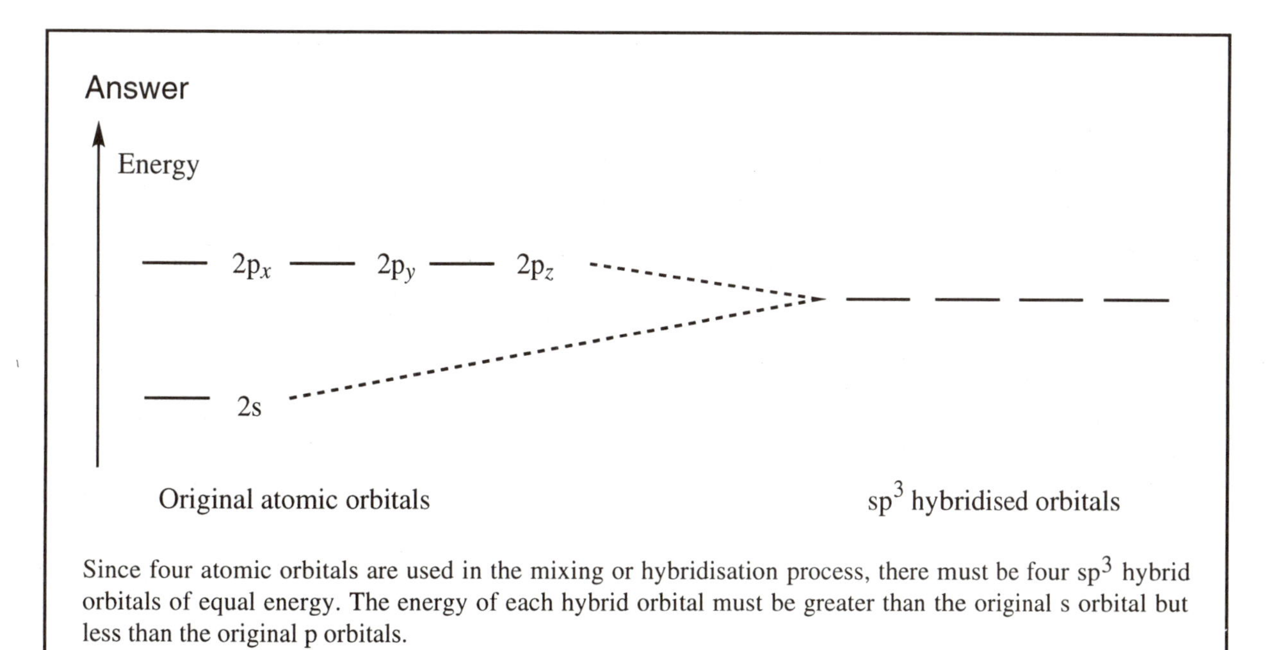

Since four atomic orbitals are used in the mixing or hybridisation process, there must be four sp^3 hybrid orbitals of equal energy. The energy of each hybrid orbital must be greater than the original s orbital but less than the original p orbitals.

We now have to fit in the valence electrons into the sp^3 hybridised orbitals. How many electrons were in the original 2s and 2p orbitals and how were they arranged? How are they now arranged in the hybridised orbitals? Draw an energy level diagram to represent this and then predict the number of bonds which carbon can make.

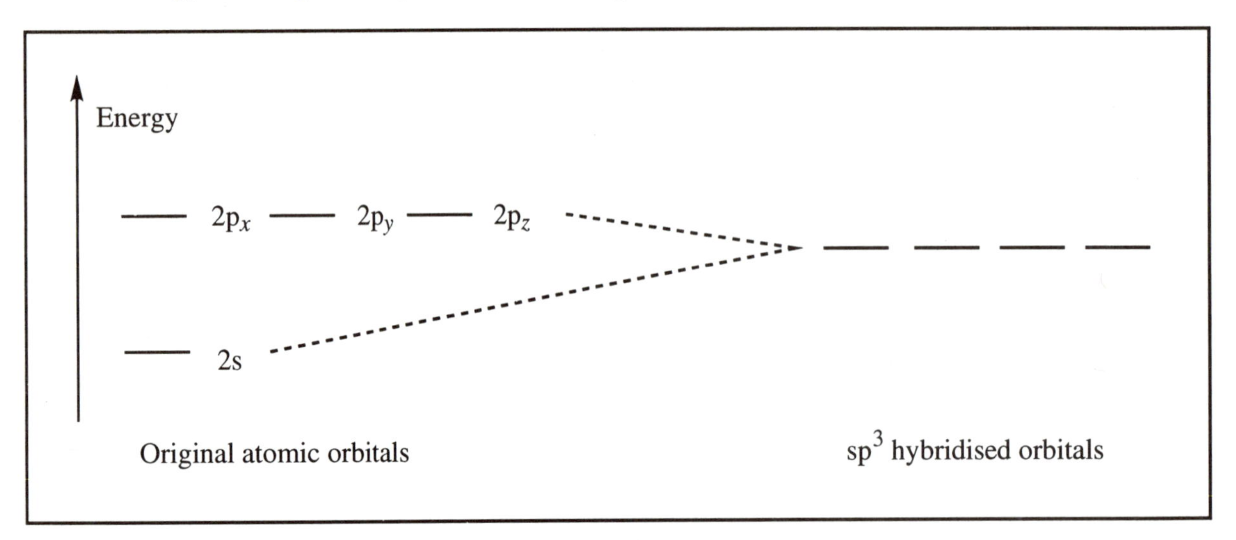

Answer

There was a total of four electrons in the original 2s and 2p orbitals. The s orbital was filled and two of the p orbitals were half-filled. There are four hybridised sp^3 orbitals all of equal energy. By Hund's rule they are all half-filled with electrons which means that there are four unpaired electrons. Four bonds are now possible

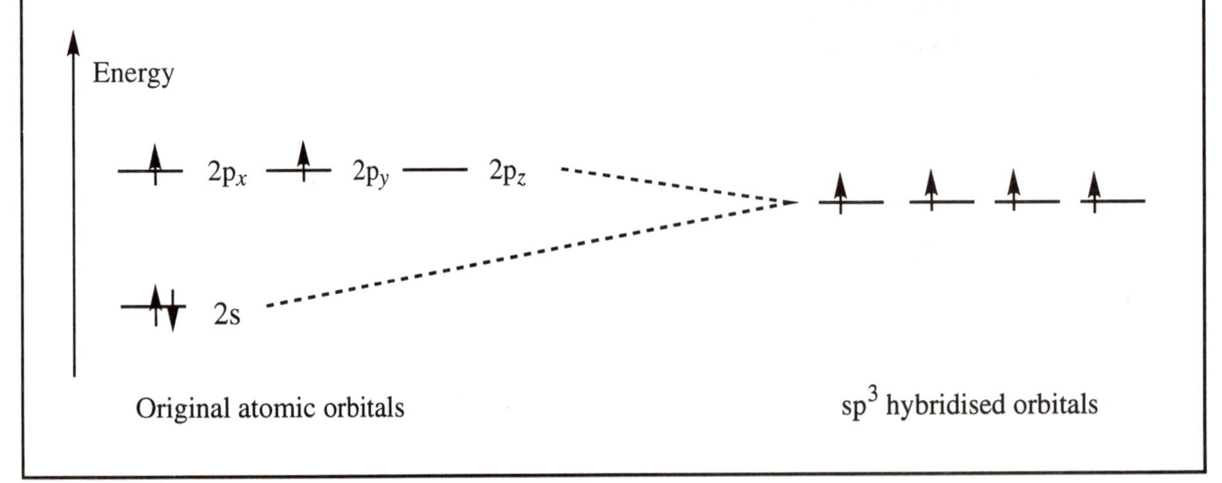

We have seen how this mixing process gives four hybridised orbitals of equal energy, but what is the shape of these orbitals? You could make a stab at predicting what that shape might be by considering the fact that there are three dumb-bell shaped 2p orbitals being mixed with one spherical shaped s orbital.

Answer

Each of the sp^3 hybridised orbitals will have the same shape and that shape will be more like a p orbital than an s orbital since more p orbitals were involved in the mixing process. The actual shape of the sp^3 orbital is a rather deformed looking dumbell shape which looks as if most of the volume of one lobe (the minor lobe) has been 'squeezed' into the other lobe (the major lobe).

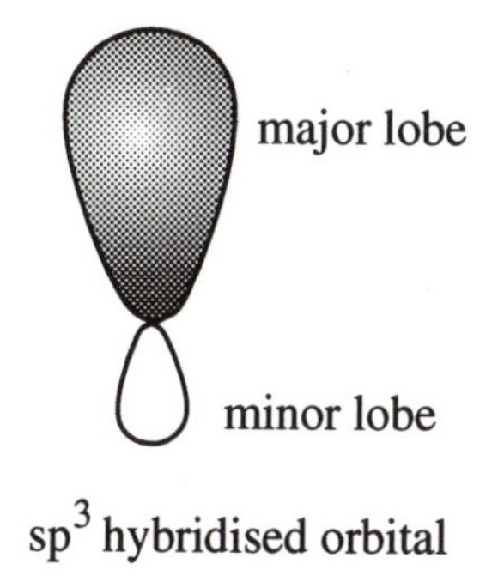

sp^3 hybridised orbital

There are four of these sp^3 hybridised orbitals. Are they likely to occupy the same area of space around the carbon nucleus?

Answer

No. Each sp^3 orbital will try to occupy a different area of space and the orbitals will try to occupy a space as far apart from each other as possible.

What sort of shape will the four hybridised orbitals form in order to be as far apart from each other as possible? (Consider only the major lobes of each orbital.) You can try out an experiment with two sausage-shaped balloons. If you tie them both together in the middle, you will end up with four lobes representing the major lobes of the four hybridised orbitals. The lobes will be as far apart from each other as possible.

Answer

If you made the balloon model, you would find that the four lobes can get as far apart from each other as possible by pointing to the corners of a tetrahedron. The angle between each of these lobes is 109°.

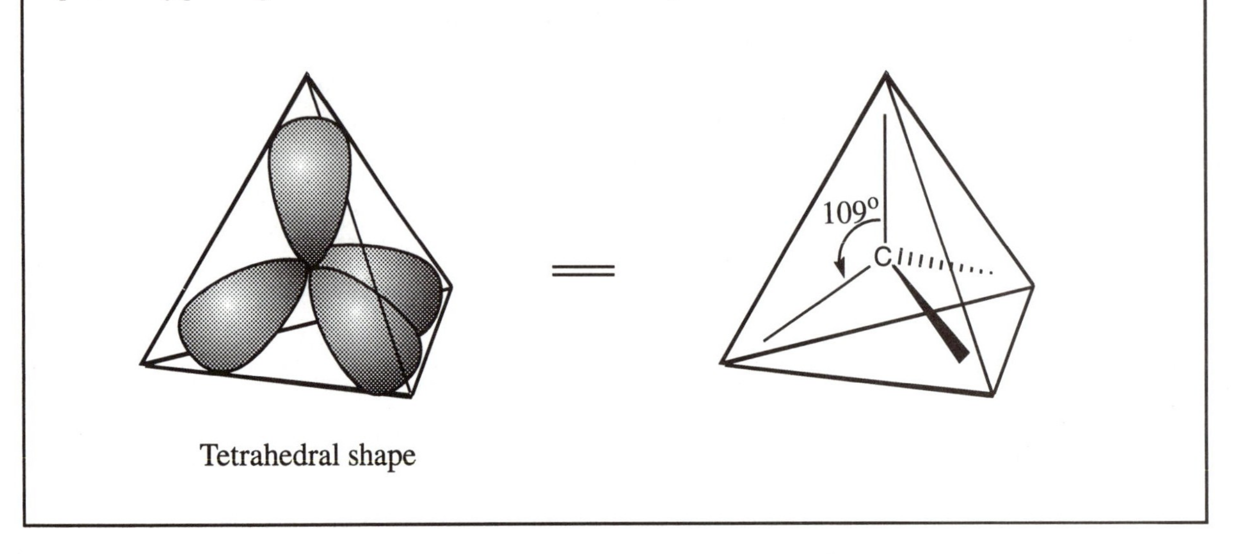

Tetrahedral shape

Hybridisation of a 2s orbital with three 2p orbitals gives four half-filled sp^3 hybridised orbitals which form the shape of a tetrahedron. Thus, the theory of hybridisation explains how carbon can form four bonds and also explains why the carbon atom is tetrahedral in saturated organic molecules such as alkanes.

Each of these hybridised orbitals can be used for a bond. When two sp^3 hybridised carbon atoms bond to each other, they each use one of their half-filled hybridised orbitals.

Draw a diagram to show this bonding, showing only the sp^3 orbitals which are bonding.

Answer

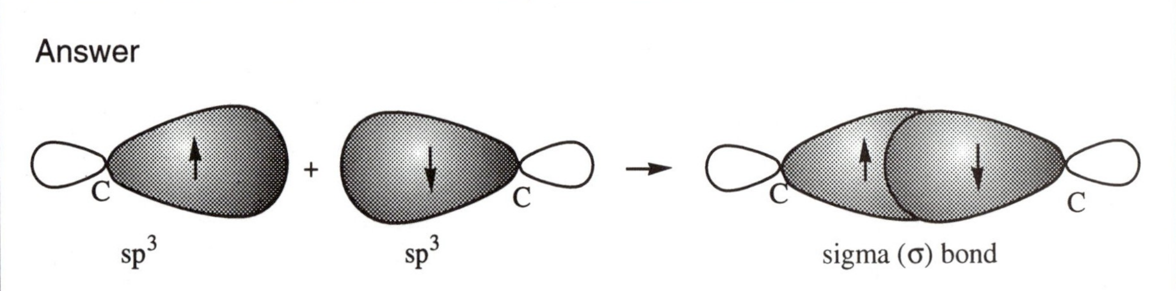

In the diagram the major lobes of the two sp^3 orbitals point towards each other and therefore have a strong interaction. The bond that results from this overlap is known as a σ bond.

What orbitals are used to form a bond between a carbon atom and a hydrogen atom? Draw a diagram to represent this.

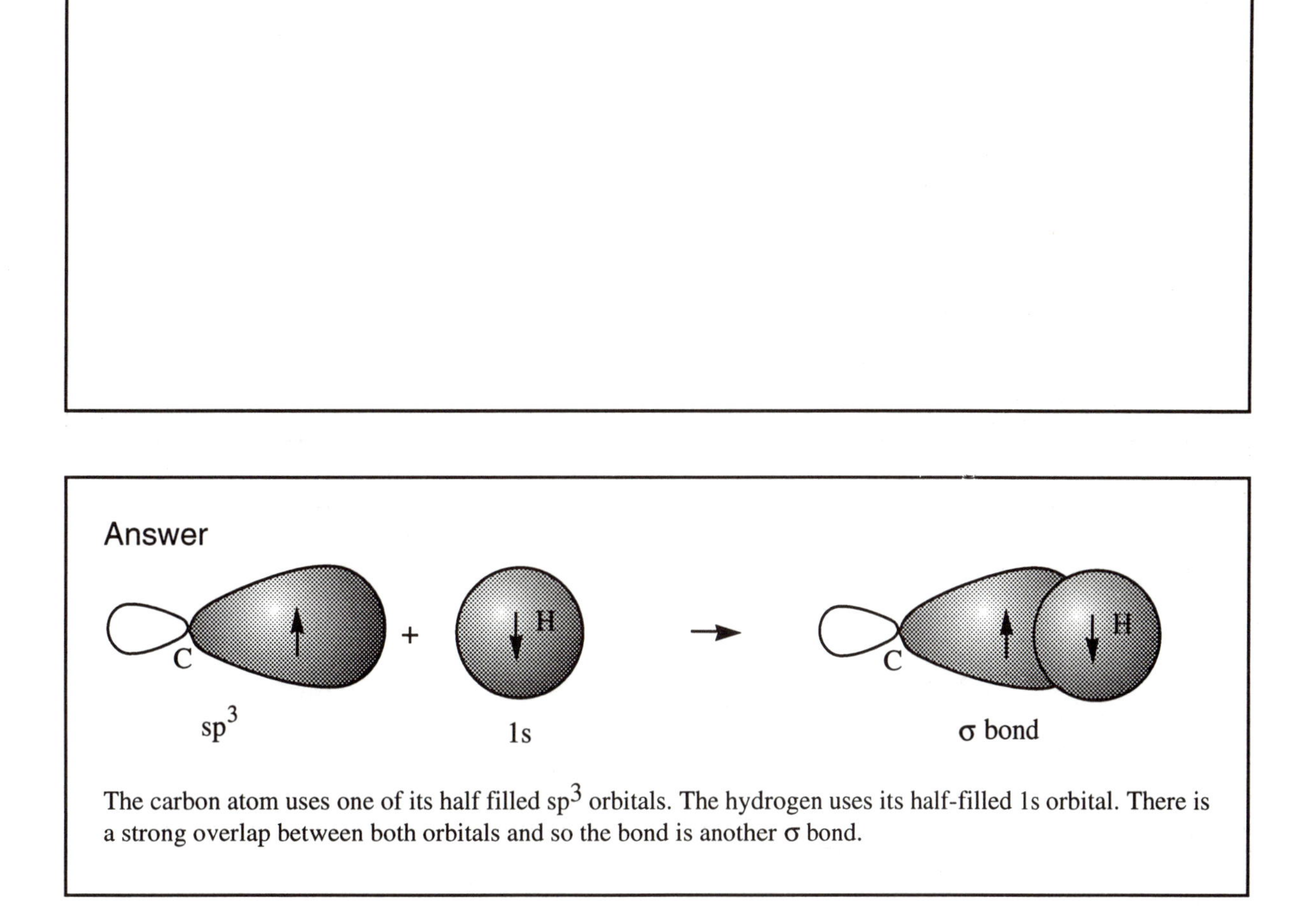

Answer

The carbon atom uses one of its half filled sp^3 orbitals. The hydrogen uses its half-filled 1s orbital. There is a strong overlap between both orbitals and so the bond is another σ bond.

14.5 sp^3 bonding with nitrogen, oxygen and chlorine

The same sort of hybridisation which takes place with carbon can also take place for nitrogen, oxygen and chlorine when these atoms are incorporated into organic molecules. Therefore, the nitrogen atom in an amine is sp^3 hybridised. The oxygen atom in an ether is sp^3 hybridised and the chlorine atom in an alkyl chloride is sp^3 hybridised.

From their positions in the periodic table, predict the number of electrons in the valence shell for each of the atoms nitrogen, oxygen and chlorine and, in the following energy level diagrams, show how these electrons will be arranged in the atomic orbitals and the sp^3 hybridised orbitals.

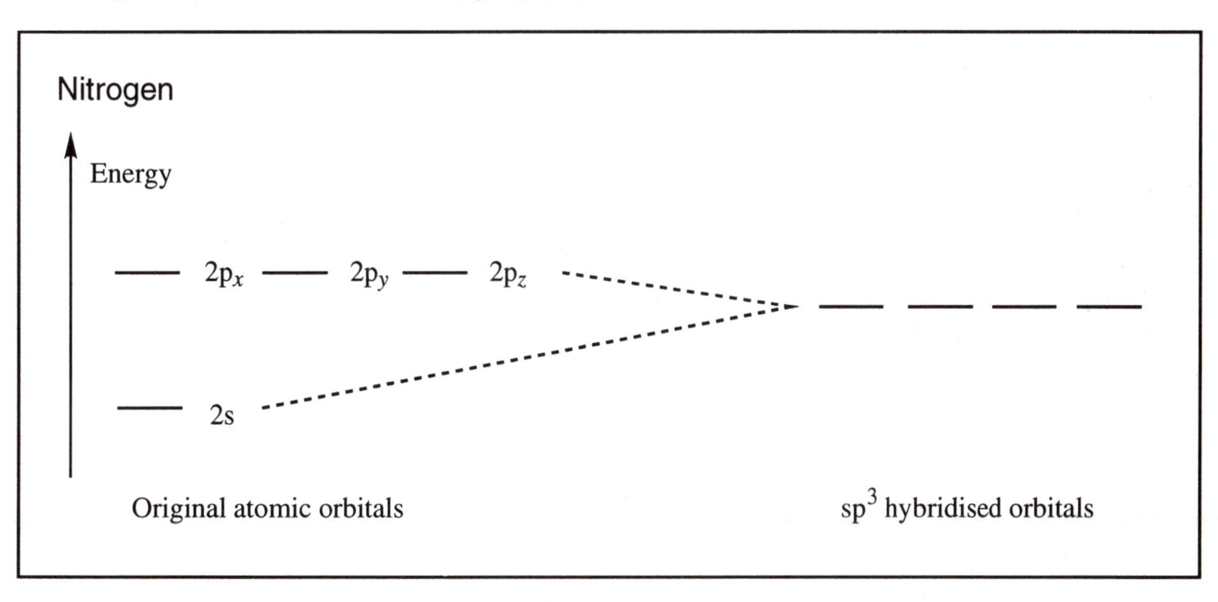

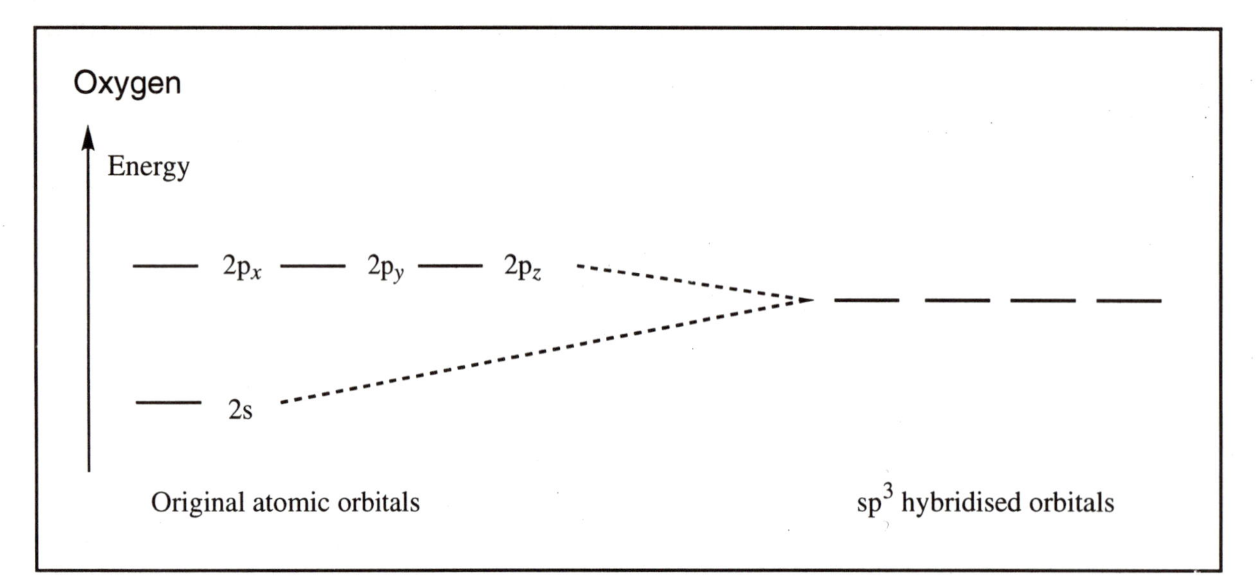

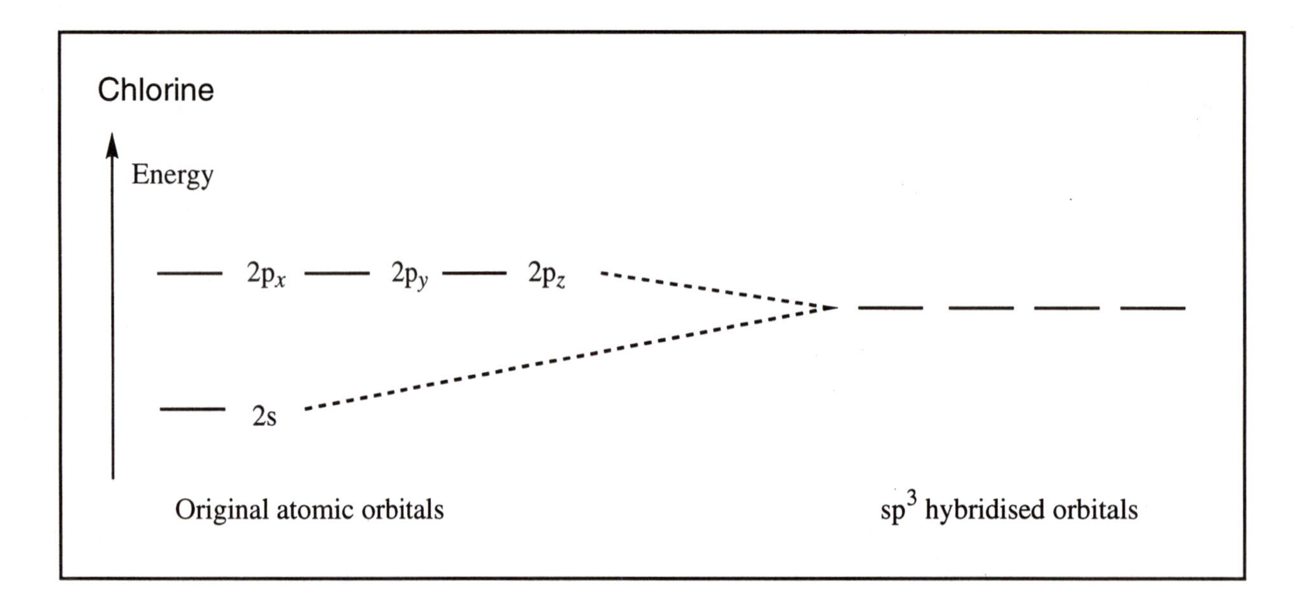

Answer

Nitrogen has five valence electrons in the second shell. After hybridisation it will have one filled and three half-filled sp^3 orbitals.

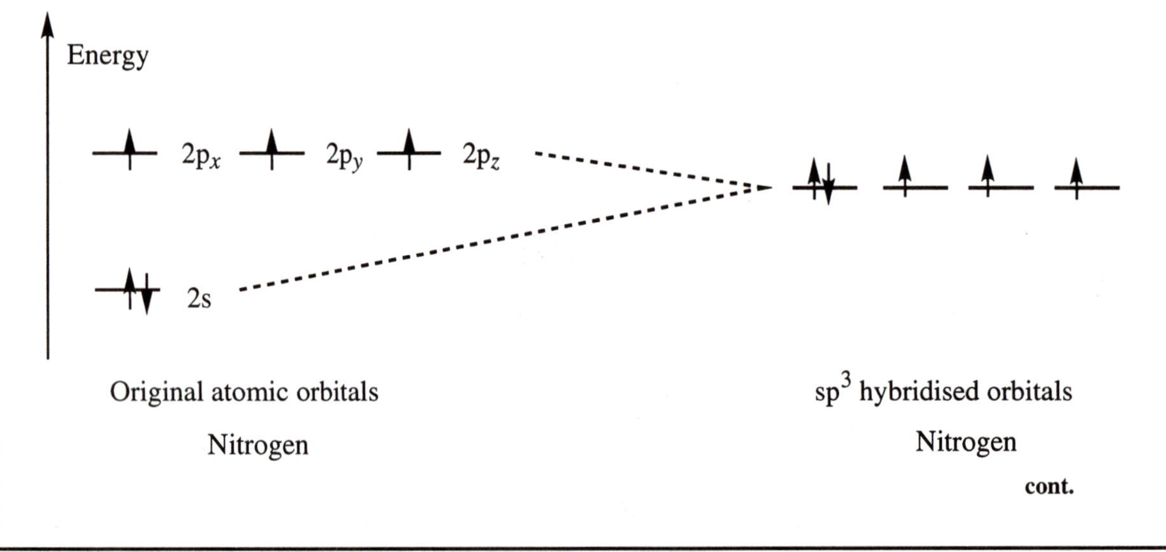

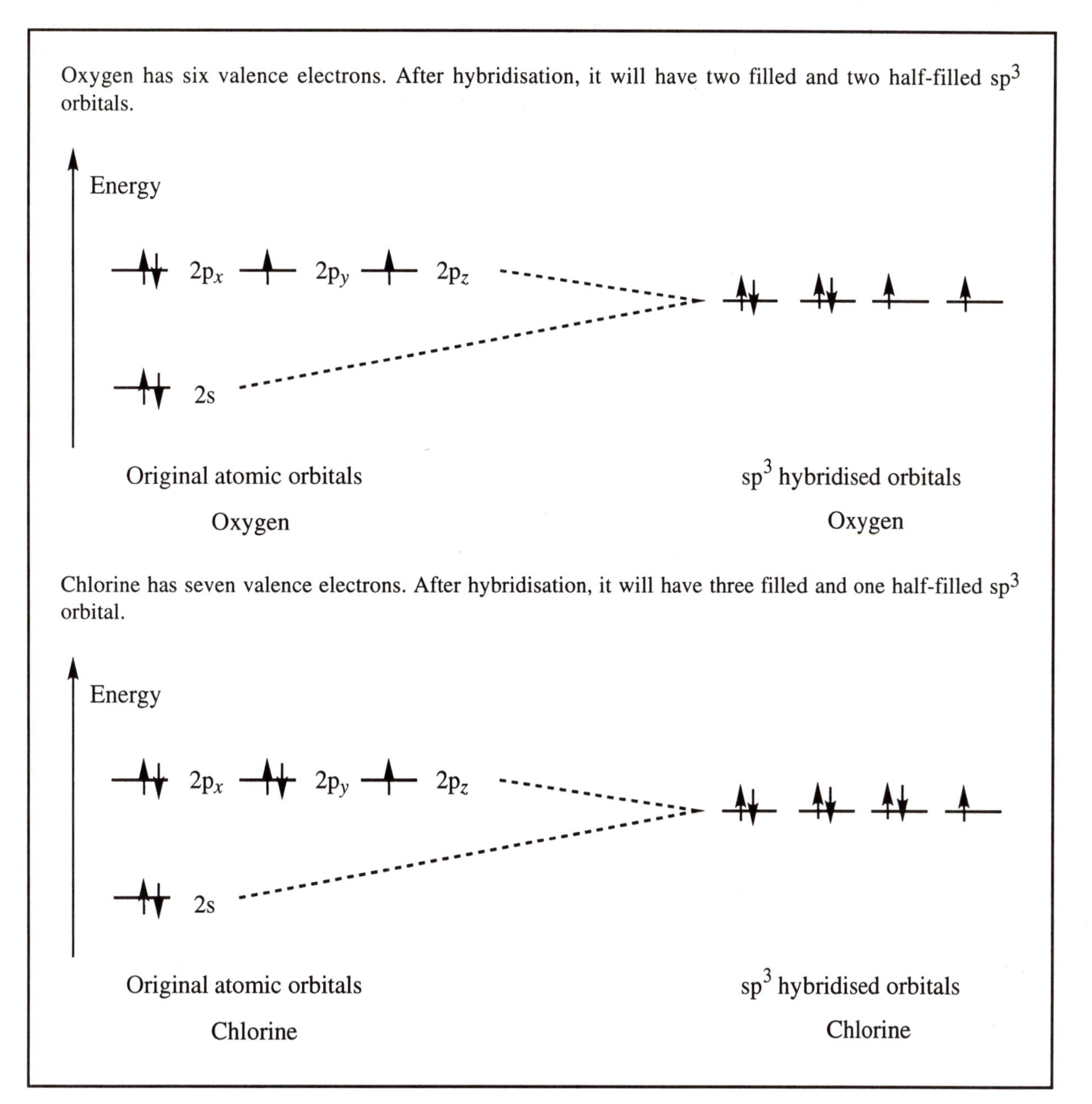
Oxygen has six valence electrons. After hybridisation, it will have two filled and two half-filled sp^3 orbitals.

Chlorine has seven valence electrons. After hybridisation, it will have three filled and one half-filled sp^3 orbital.

From the above diagrams, predict the number of bonds which each of these atoms can form in organic molecules.

Answer

The number of possible bonds equals the number of half-filled sp^3 orbitals. Therefore, nitrogen can form three bonds, oxygen two, and chlorine one.

Since nitrogen is sp^3 hybridised, what sort of shape would you expect methylamine to have at the nitrogen atom? Draw the structure.

Answer

The four sp^3 orbitals will form a tetrahedral arrangement with one of the orbitals occupied by a lone pair of electrons. If one just considers the bonds, the shape will be pyramidal.

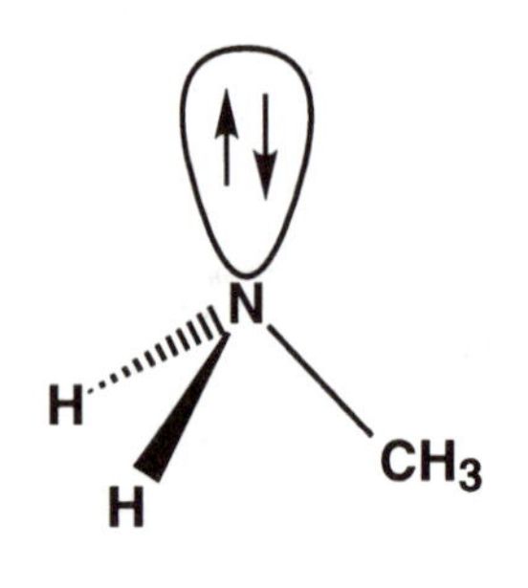

Since oxygen is sp^3 hybridised, what sort of shape would you expect methanol to have at the oxygen atom? Draw the structure with the lone pairs included.

Answer

The four sp^3 orbitals will form a tetrahedral arrangement with two of the orbitals occupied by lone pairs of electrons. If one just considers the bonds, the shape will be angled or bent.

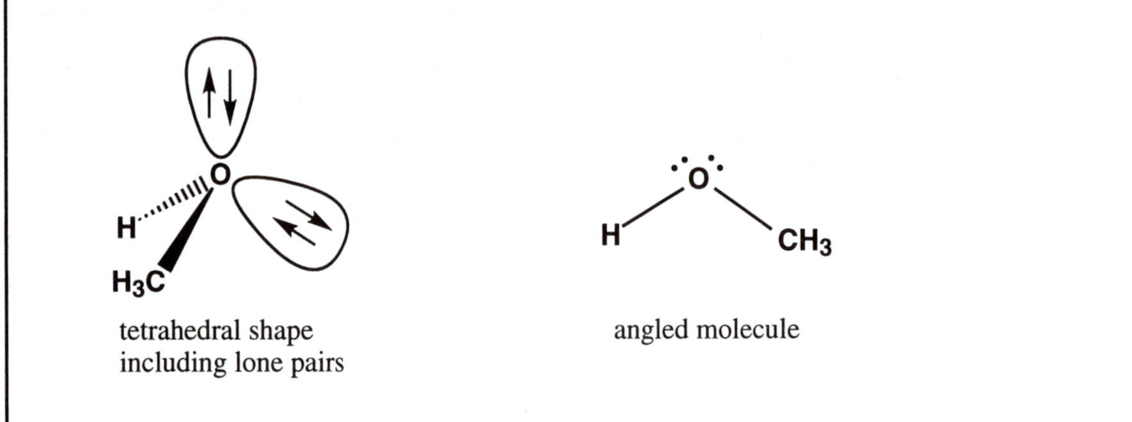

What sorts of bonds are present in the following classes of compounds—alkanes, alcohols, amines, alkyl halides and ethers?

Answer

All these structures contain single bonds between atoms. The bonds are all σ bonds.

Identify the atoms which are sp^3 hybridised in the following molecules.

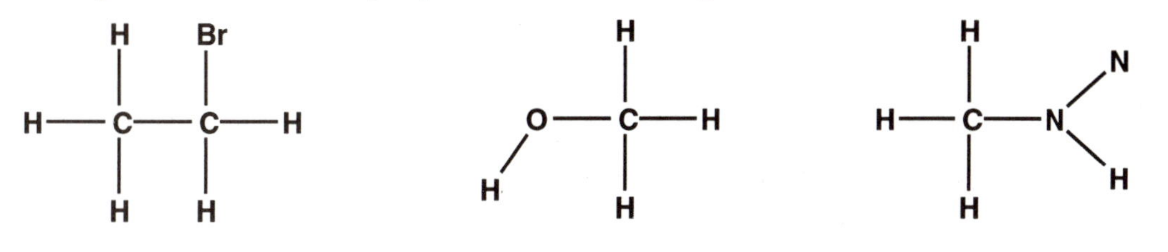

Answer

The only atoms which are not sp^3 hybridised are the hydrogen atoms. They form bonds using their 1s orbitals.

14.6 sp^2 hybridisation

In sp^3 hybridisation, the 2s orbital is mixed with all three of the 2p orbitals to give a set of hybrid orbitals. In sp^2 hybridisation, the s orbital is mixed with two of the 2p orbitals (e.g. $2p_x$ and $2p_z$) and the remaining $2p_y$ orbital is unaffected. How many sp^2 hybrid orbitals are formed?

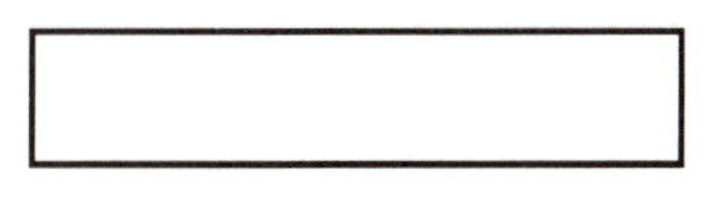

Answer

Three atomic orbitals were used in the hybridisation process. Therefore, there must be three hybrid sp^2 orbitals resulting. The number of hybrid orbitals must equal the number of original atomic orbitals used for mixing.

The hybrid orbitals will each have the same energy but be different in energy from the original atomic orbitals. That energy difference will reflect the mixing of the respective atomic orbitals. Draw an energy diagram comparing the energies of the original atomic orbitals, the resulting sp^2 hybridised orbitals and the remaining $2p_y$ orbital.

Energy

Original atomic orbitals

sp^2 hybridised orbitals

Answer

Since three atomic orbitals are used in the mixing or hybridisation process, there must be three sp^2 hybrid orbitals of equal energy. The energy of each hybrid orbital must be greater than the original s orbital but less than the original p orbitals. The remaining 2p orbital remains at its original energy level. In the diagram, this is the $2p_y$ orbital.

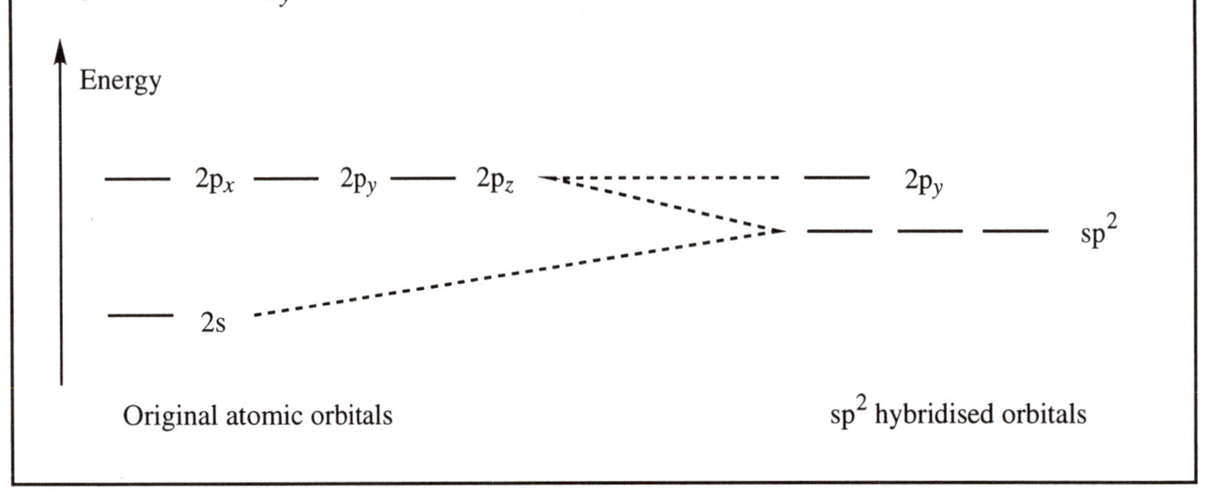

This diagram is relevant for carbon, oxygen and nitrogen, all of which can be sp^2 hybridised. We shall concentrate on carbon and look at how the valence electrons of carbon are fitted into the three sp^2 hybridised orbitals and the remaining $2p_y$ orbital. Place the electrons into the energy level diagram below, making the assumption that the energy difference between the sp^2 orbitals and the remaining $2p_y$ orbital is small.

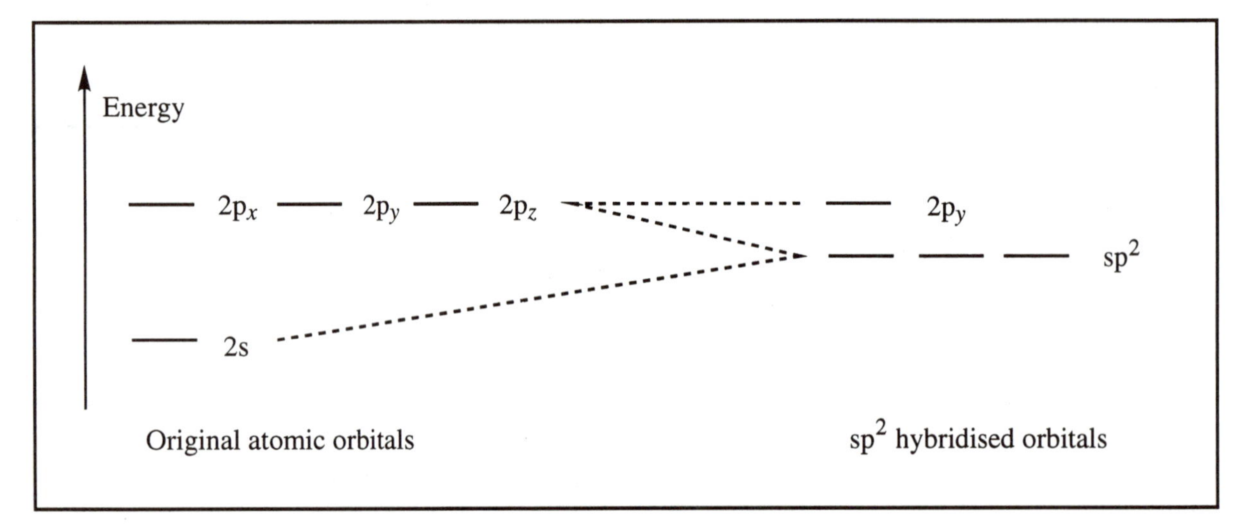

Answer

There are three hybridised sp^2 orbitals, all of equal energy. The first three electrons are fitted into each one according to Hund's rule such that they are all half-filled with electrons. This leaves one electron left over. You have a choice between pairing it up in a half-filled sp^2 orbital or placing it into the vacant $2p_y$ orbital. The usual principle is to fill up orbitals of equal energy before moving to orbitals of higher energy. However, if the energy difference between orbitals is small (as in this case) it is easier for the electrons to fit into the higher energy orbitals rather than to pair up. Therefore, the last electron in this case fits into the $2p_y$ orbital, resulting in three half-filled sp^2 orbitals, and one half-filled $2p_y$ orbital.

cont.

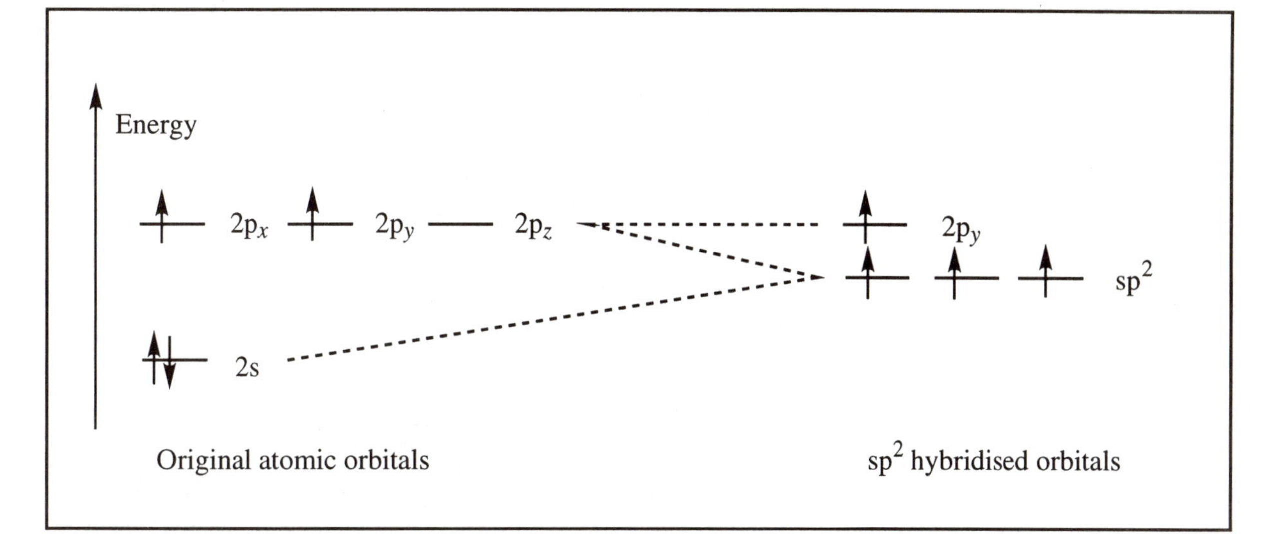

How many bonds are possible for an sp^2 hybridised carbon?

Answer

There are four half-filled orbitals and so you would expect four bonds.

The $2p_y$ orbital will have the same dumb-bell shape as before. What shapes would you expect for the sp^2 orbitals?

Answer

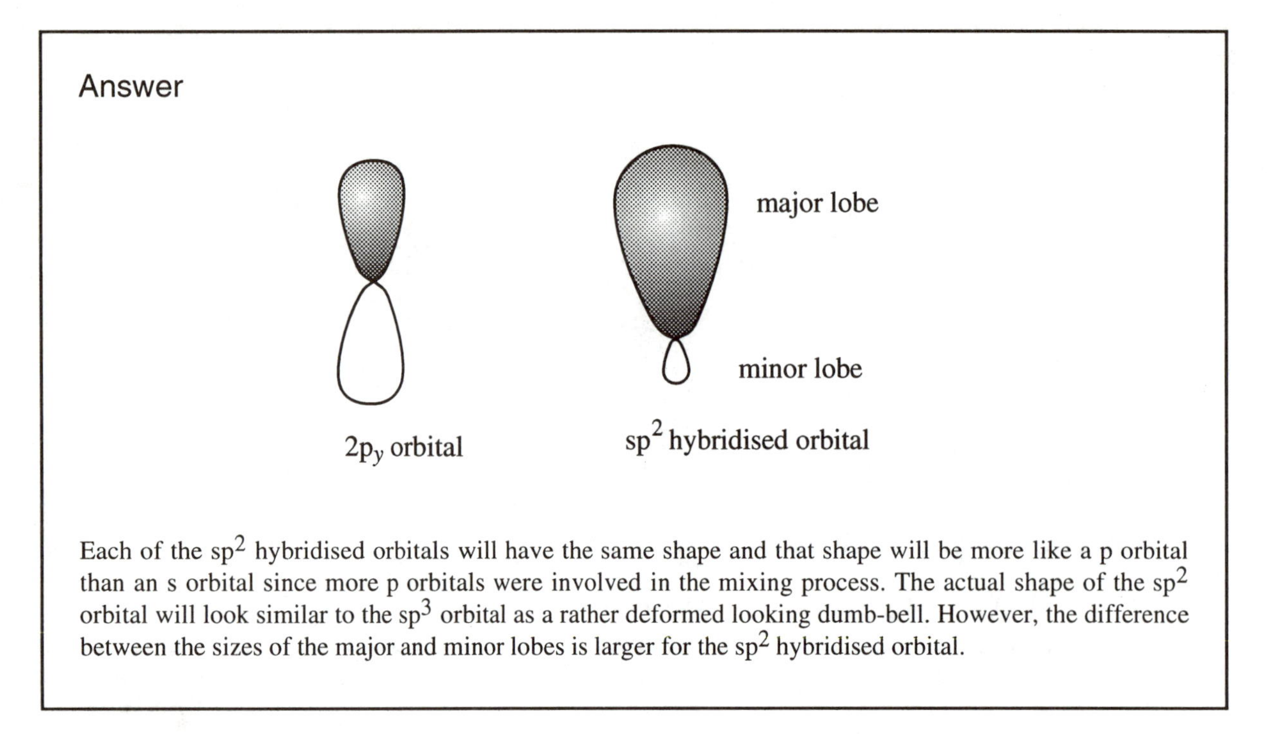

Each of the sp^2 hybridised orbitals will have the same shape and that shape will be more like a p orbital than an s orbital since more p orbitals were involved in the mixing process. The actual shape of the sp^2 orbital will look similar to the sp^3 orbital as a rather deformed looking dumb-bell. However, the difference between the sizes of the major and minor lobes is larger for the sp^2 hybridised orbital.

There are three sp^2 hybridised orbitals and one 2p$_y$ orbital. Each orbital will try to occupy a different space as far apart from each other as possible.

Draw the 2p$_y$ orbital in the following diagram.

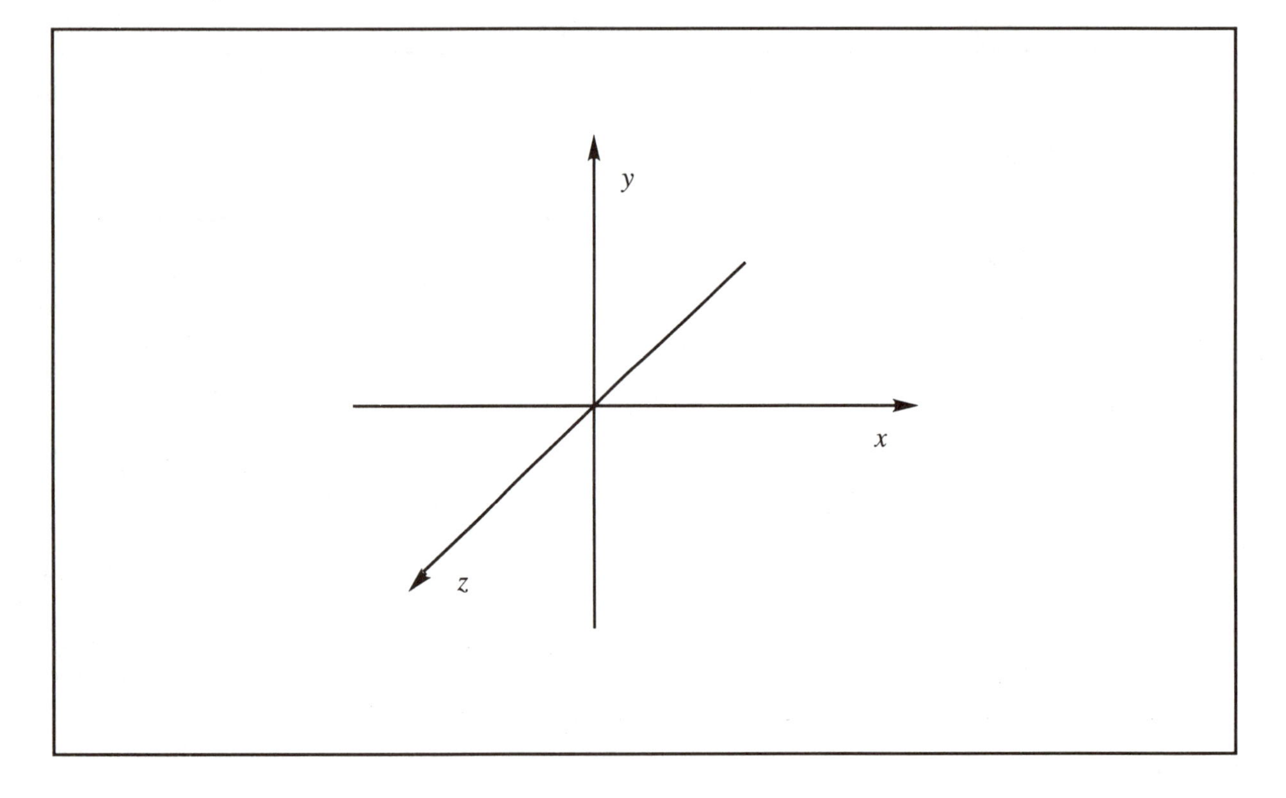

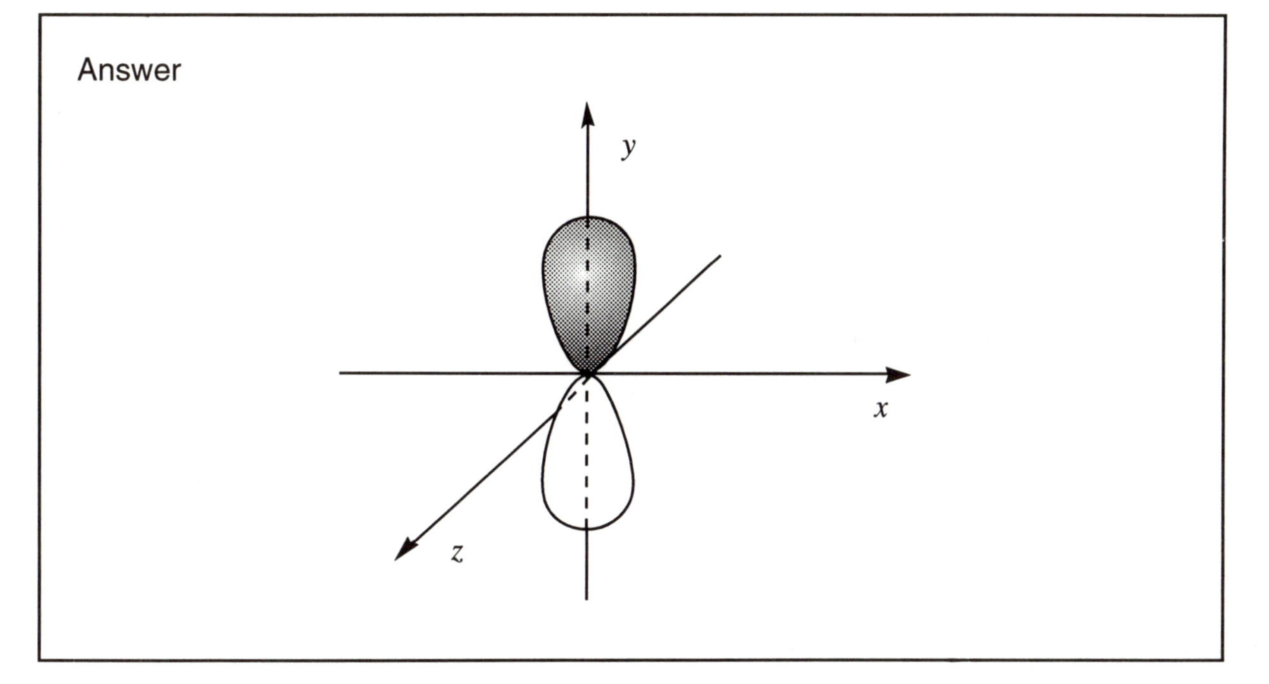

The lobes of the $2p_y$ orbital occupy the space above and below the plane of the x and z axes. Taking this into account draw in the three sp^2 orbitals (major lobes only) such that they are as far apart from the $2p_y$ orbital and each other as possible.

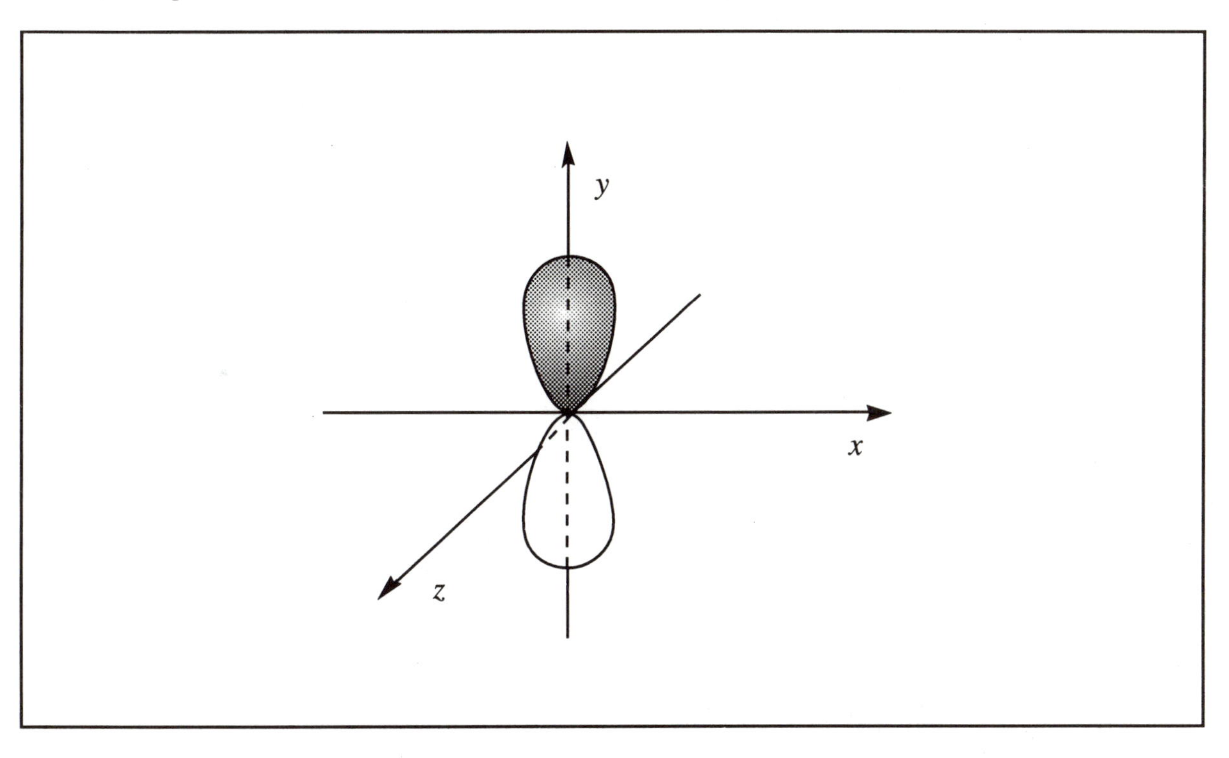

Answer

They will all be placed in the *x*–*z* plane pointing toward the corners of a triangle. In the diagram below, one sp^2 orbital is along the *x* axis, one is halfway between the negative *x* axis and the *z* axis, and the third (partly hidden) is halfway between the negative *x* and negative *z* axes. The angle between each of these lobes is 120°.

cont.

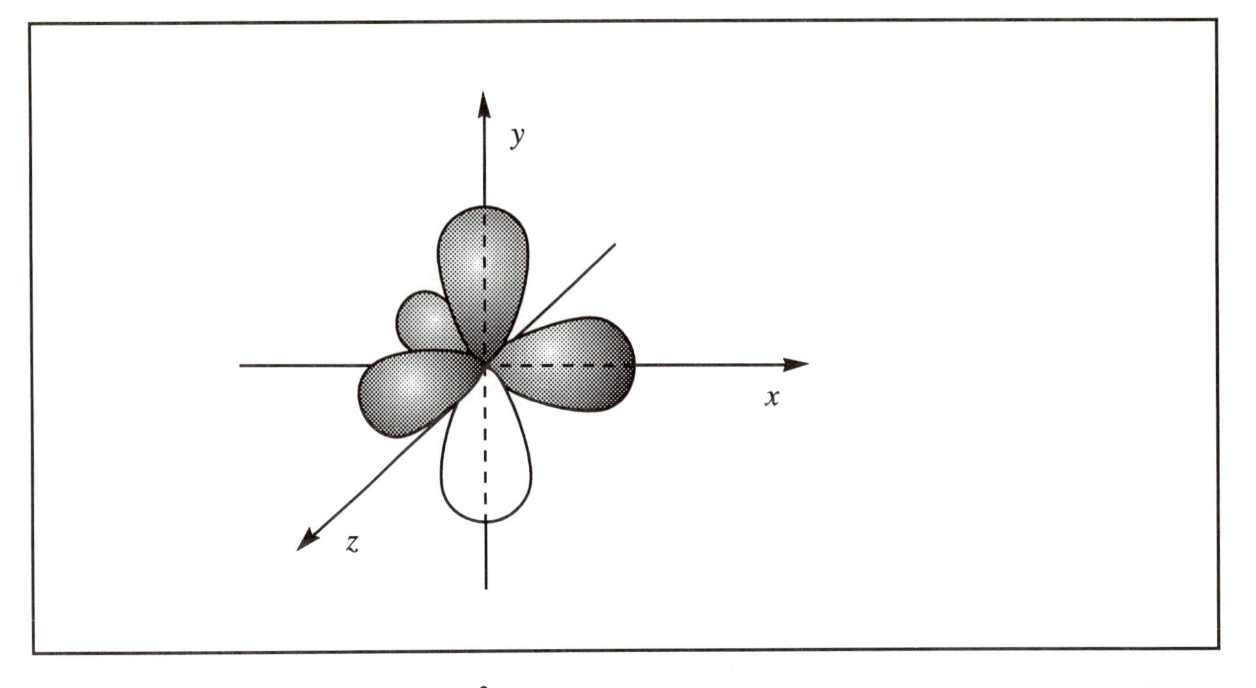

We are now ready to look at how an sp^2 hybridised carbon makes bonds in functional groups such as alkenes or carbonyl groups.

14.7 Bonding in alkenes

Hybridisation of a 2s orbital with two 2p orbitals gives three half-filled sp^2 hybridised orbitals which form a trigonal planar shape. If these three orbitals are used for bonding then we can explain the shape of an alkene.

Let us consider the bonding in ethene where each carbon is sp^2 hybridised. What orbitals will be used in the formation of a C–H bond?

Answer

The hydrogen atom uses a half-filled 1s orbital to bond with a half-filled sp^2 orbital on carbon.

Draw a diagram to represent this.

Answer

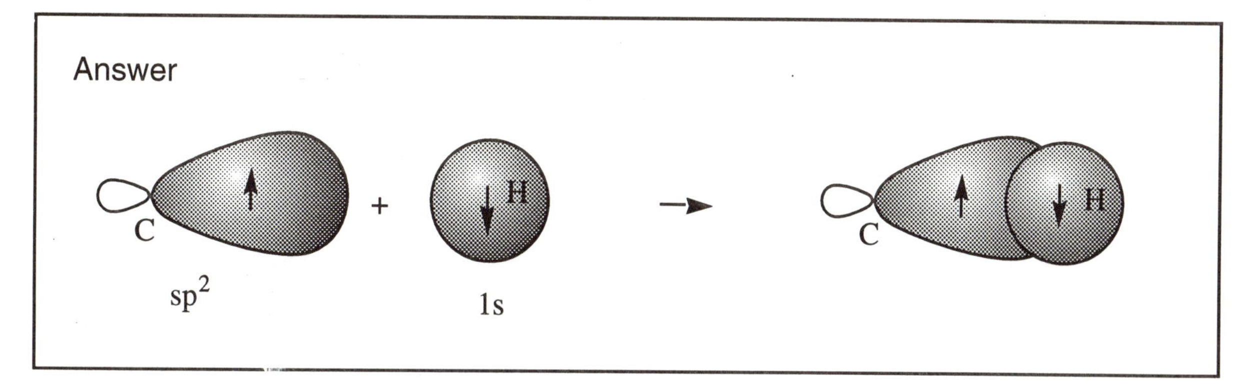

What sort of bond is this? Is it strong or weak and why?

Answer

The C–H bond is a s bond which is a strong bond due to good overlap between the bonding orbitals.

Draw a diagram to show the formation of a σ bond between two sp^2 hybridised carbon atoms.

Answer

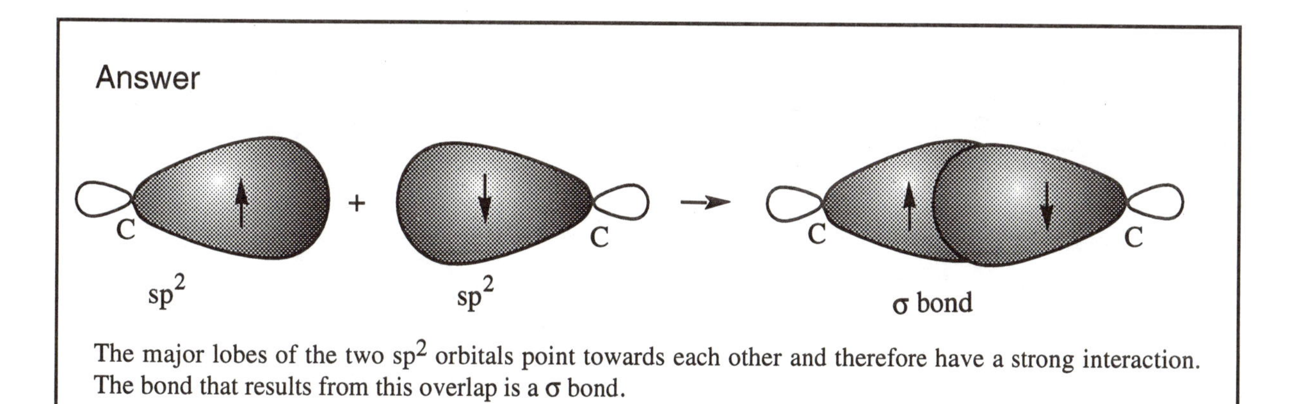

The major lobes of the two sp^2 orbitals point towards each other and therefore have a strong interaction. The bond that results from this overlap is a σ bond.

The full σ bonding diagram for ethene is now as shown.

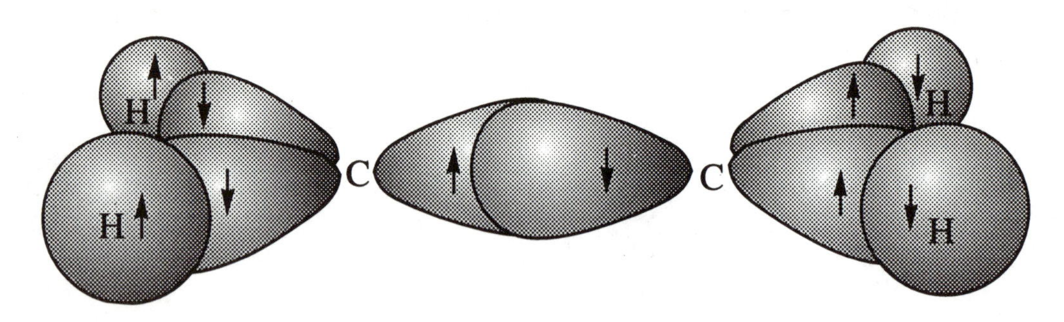

Each bond is a strong σ bond and we can simplify the diagram by drawing it as follows.

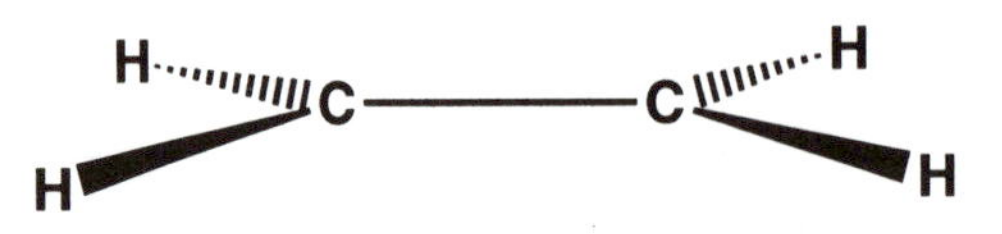

We know that ethene is a rigid, flat molecule where each carbon is trigonal planar. We have seen how the carbon atoms are trigonal planar, but we have not explained why the molecule should be rigid and planar overall. If the σ bonds were the only bonds present in ethene, would you expect the molecule to remain planar as drawn?

Answer

If the σ bonds were the only bonds present in ethene, the molecule would not remain planar since rotation could occur round the C–C σ bond.

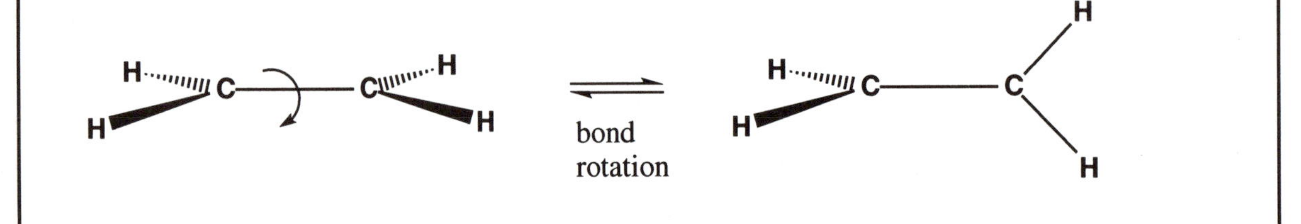

There must be some other sort of bonding which 'locks' the alkene into this planar shape. This bond must clearly involve the remaining half-filled $2p_y$ orbitals on each carbon. Draw in the half-filled $2p_y$ orbitals in the following diagram.

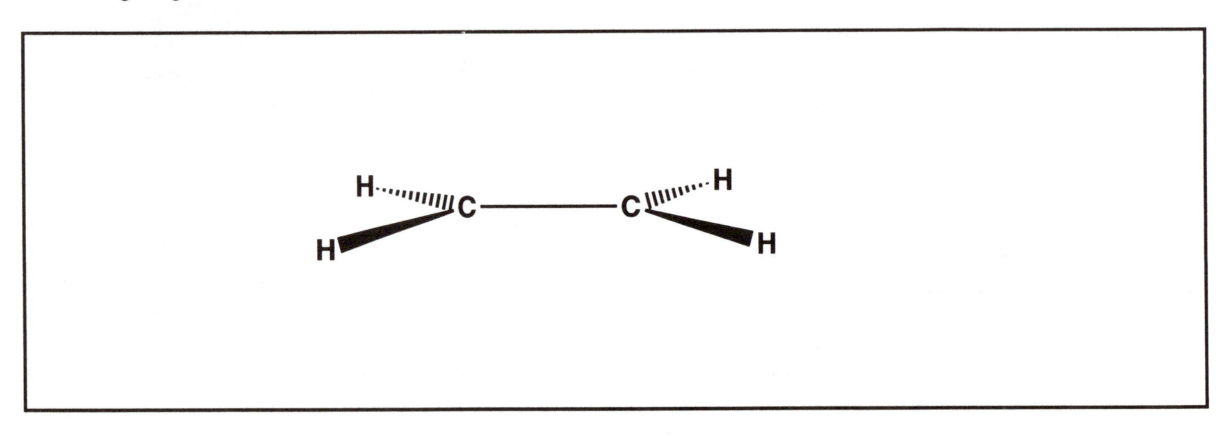

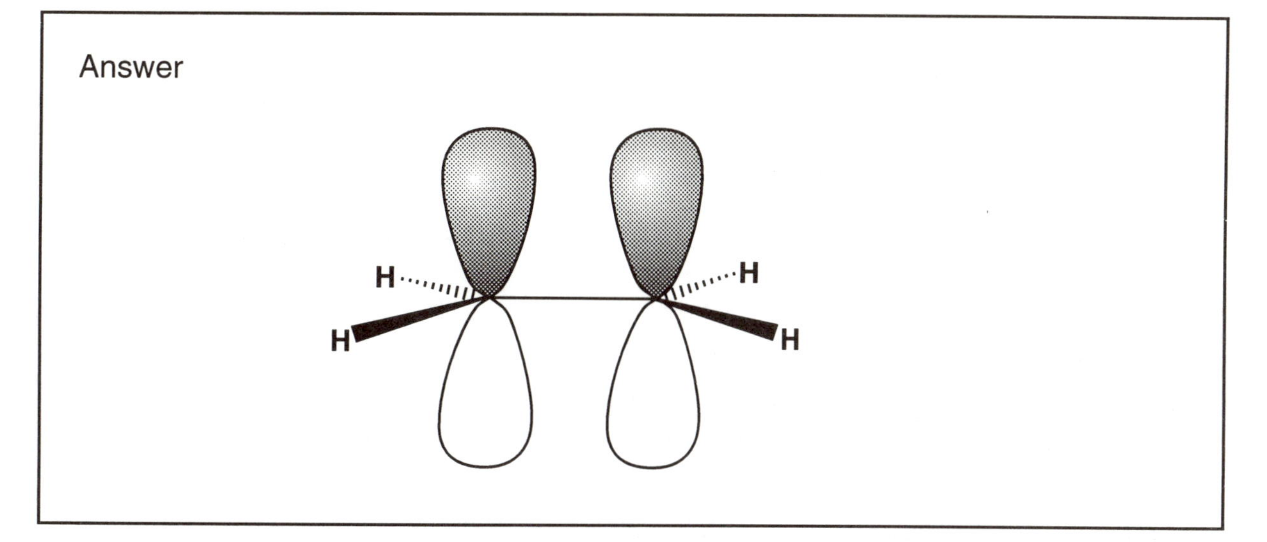

How could these $2p_y$ orbitals be used to form another bond between the carbon atoms? Demonstrate this on the following diagram and draw the shape of the bonding orbital which would be formed.

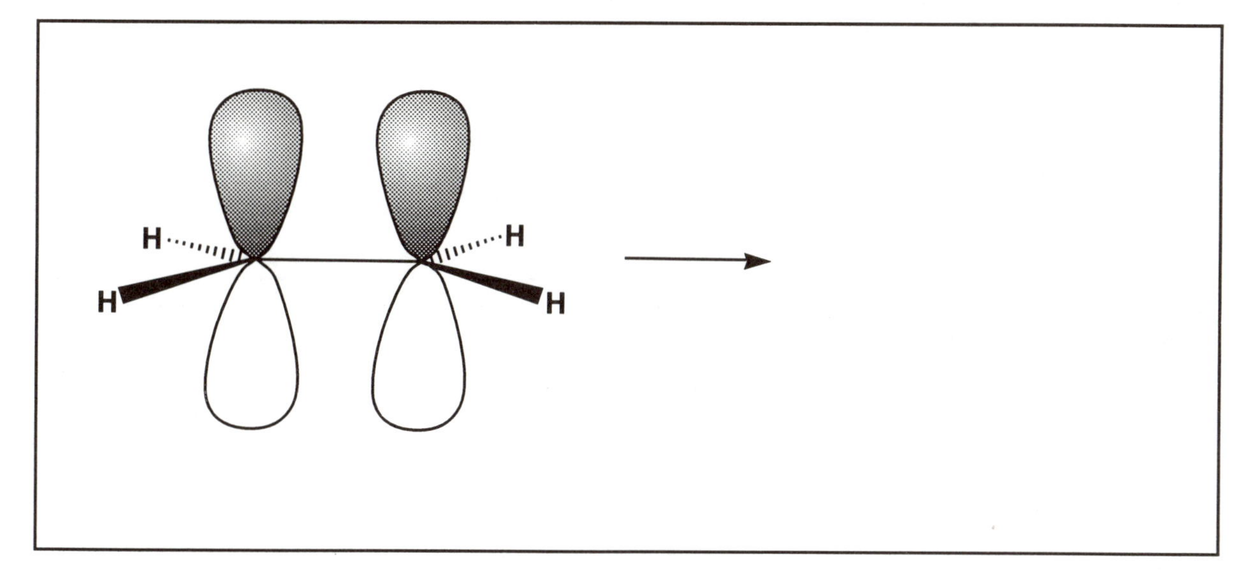

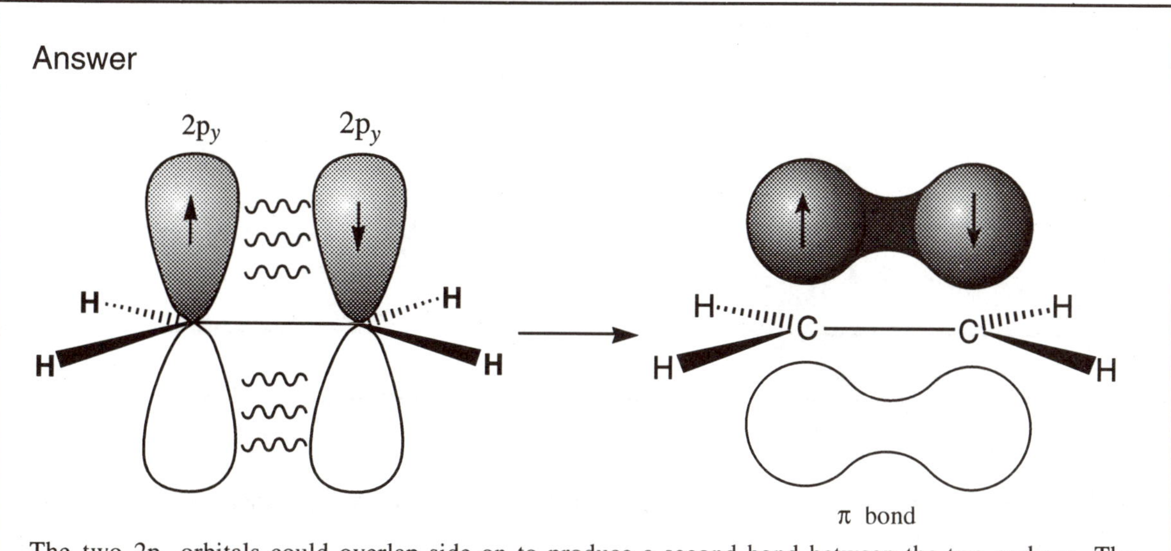

π bond

The two $2p_y$ orbitals could overlap side on to produce a second bond between the two carbons. The resulting bond (known as a pi (π) bond), would have lobes above and below the plane of the molecule.

How would a π bond force ethene to be a rigid, planar molecule?

Answer

The π bond prevents rotation round the C–C bond. If rotation took place, the constituent $2p_y$ orbitals making up the π bond would no longer overlap and the π bond would be broken. This would require energy.

Would you expect a π bond to be stronger or weaker than a σ bond and why?

Answer

You would expect the π bond to be weaker. The lobes of the $2p_y$ orbitals used to form the π bond have to overlap side on. Since the $2p_y$ orbitals do not face each other, their overlap is weaker. As a result, the π bond must be weaker.

This theory of bonding explains why alkenes such as ethene are planar in shape. It also explains why alkenes are more reactive than alkanes. Why? Remember that reactions involve the making and breaking of bonds.

Answer

Alkenes have a π bond which is weaker than a σ bond. This bond is more easily broken and is more likely to take part in reactions.

14.8 Bonding in carbonyl functional groups (C=O)

Exactly the same theory can be used to explain the bonding within a carbonyl group (C=O) where both the carbon and oxygen are sp^2 hybridised. Show in the following energy level diagram how the valence electrons of oxygen are arranged after sp^2 hybridisation.

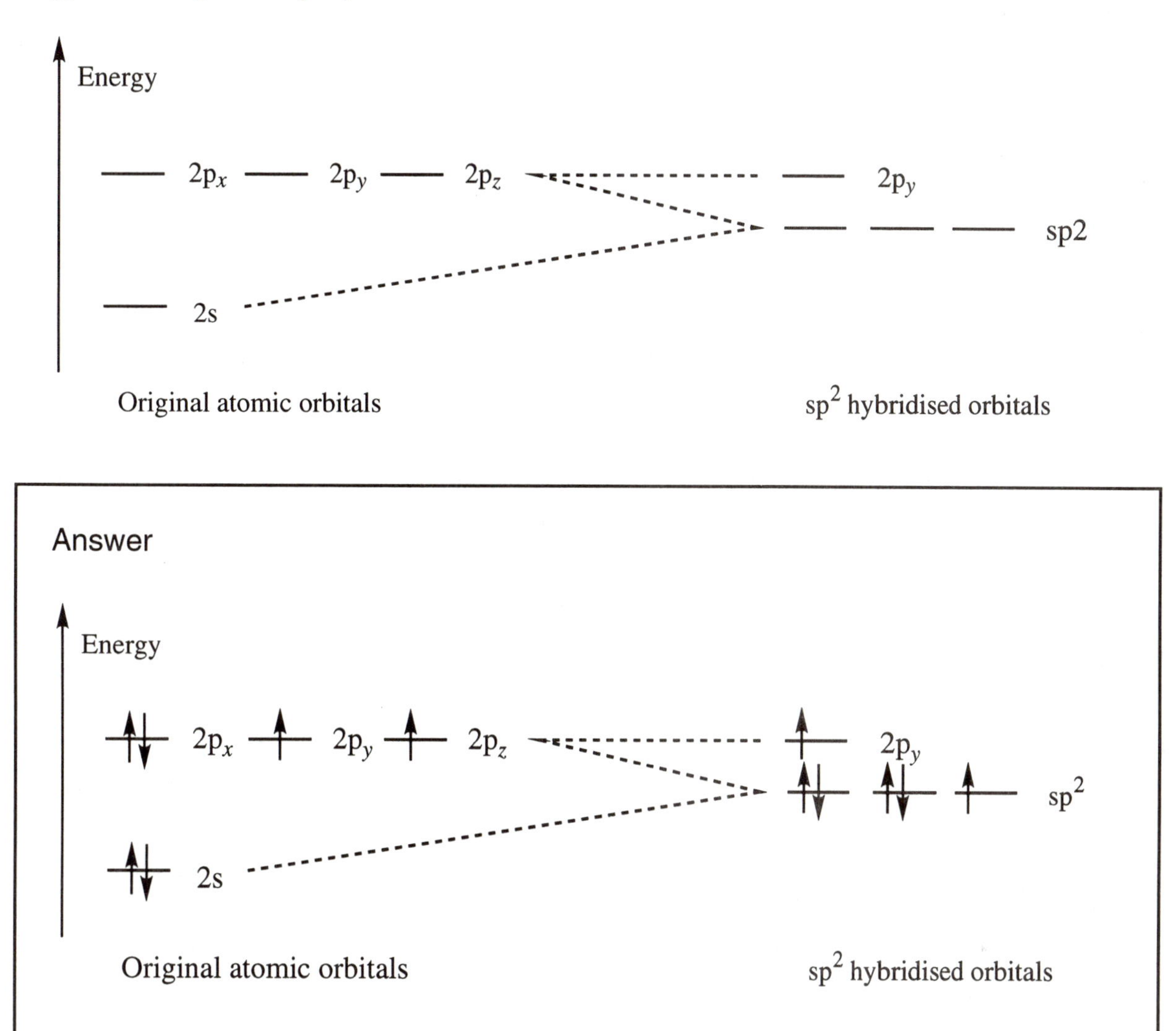

Explain how many bonds are possible for an sp^2 hybridised oxygen and whether they are identical bonds.

Answer

Two of the sp^2 hybridised orbitals are filled (lone pairs) which leaves two half-filled orbitals available for bonding. The sp^2 orbital can be used for a strong σ bond while the $2p_y$ orbital can be used for the weaker π bond.

Identify the σ and π bonds in formaldehyde.

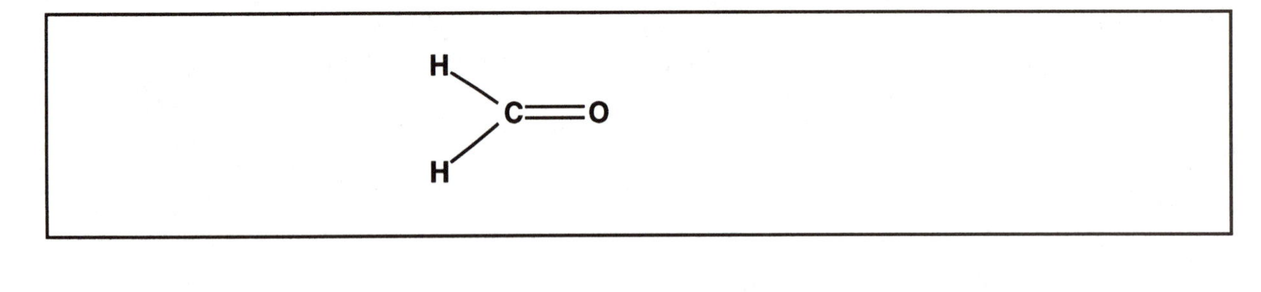

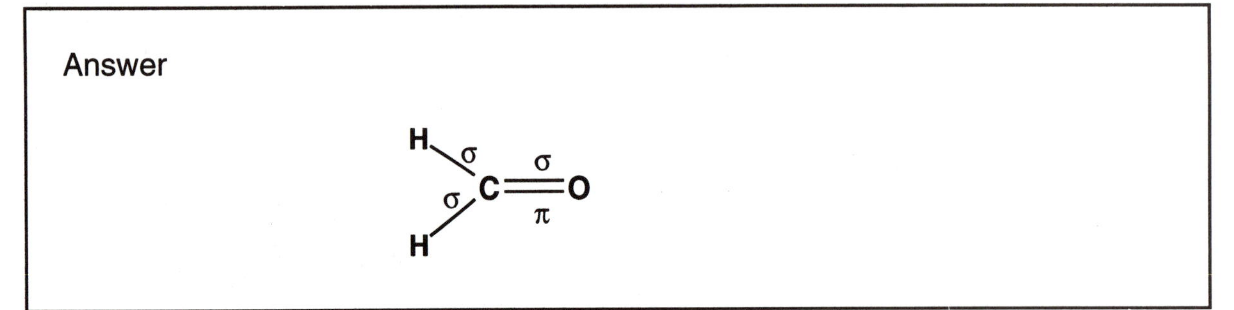

Fill in the orbitals in the following diagram to show how the σ and π bonds are formed in the carbonyl group of formaldehyde.

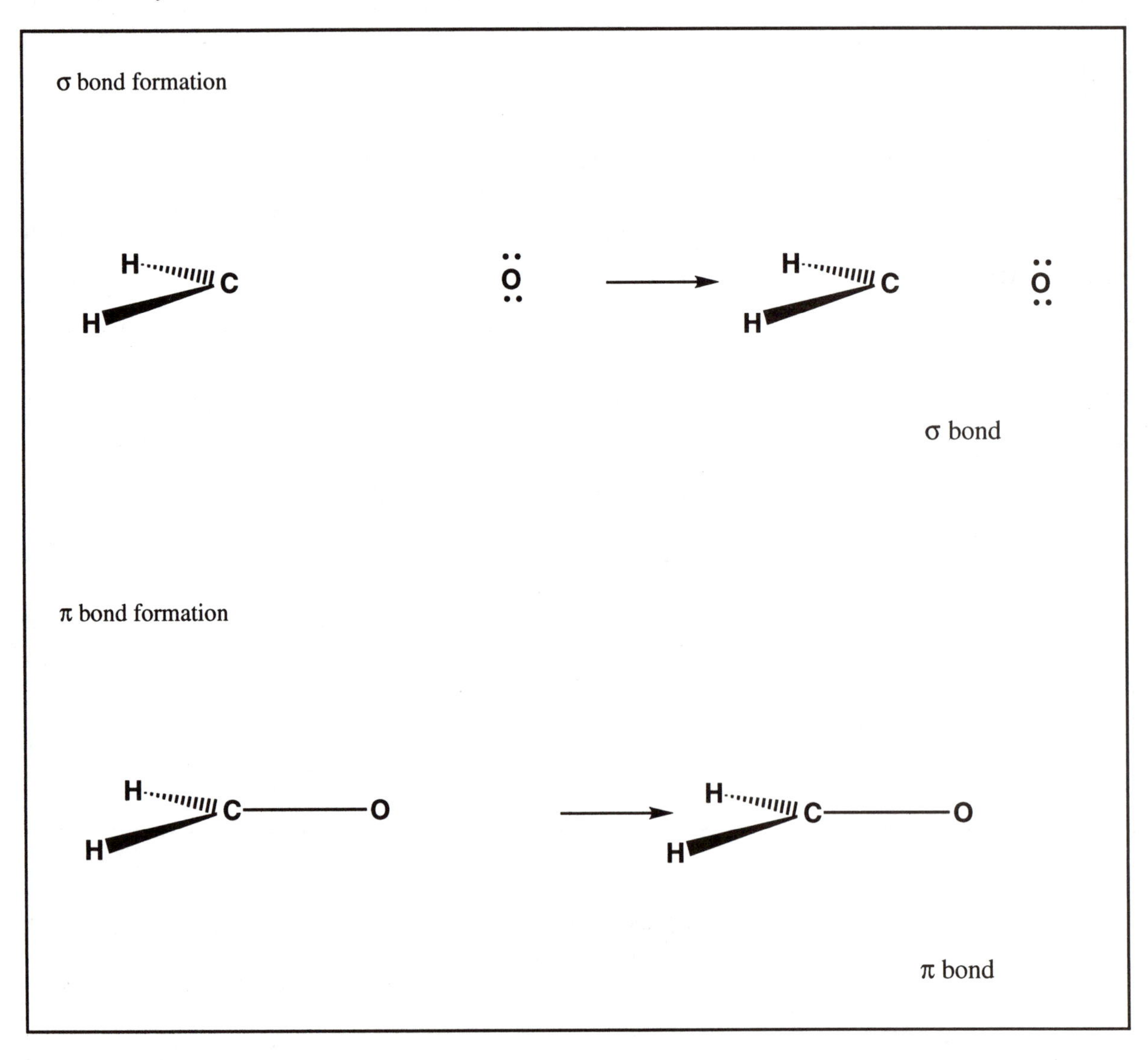

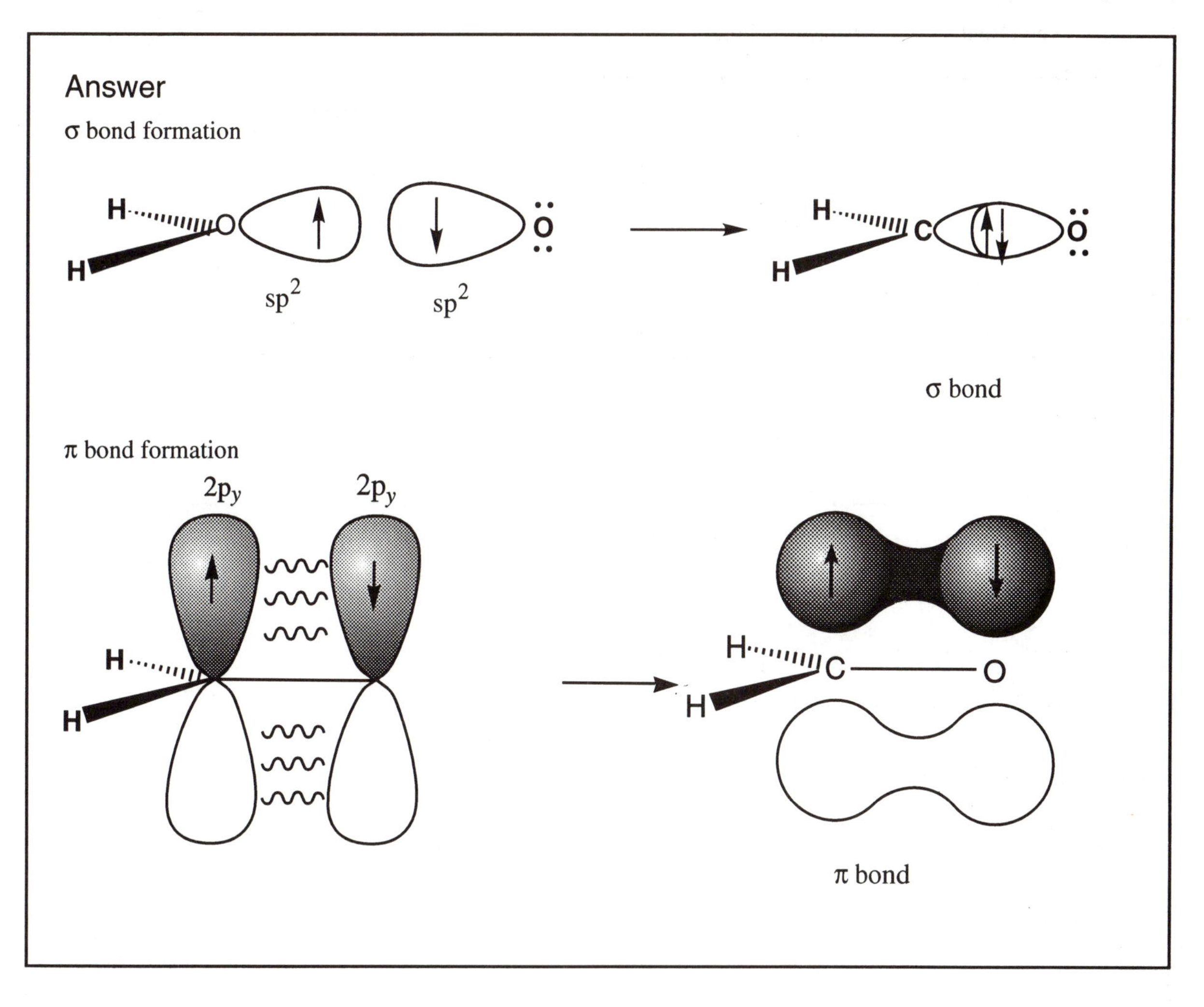

14.9 Bonding in aromatic rings

Consider the structure of benzene and identify whether each carbon atom is sp^2 or sp^3 hybridised. Identify each bond as σ or π.

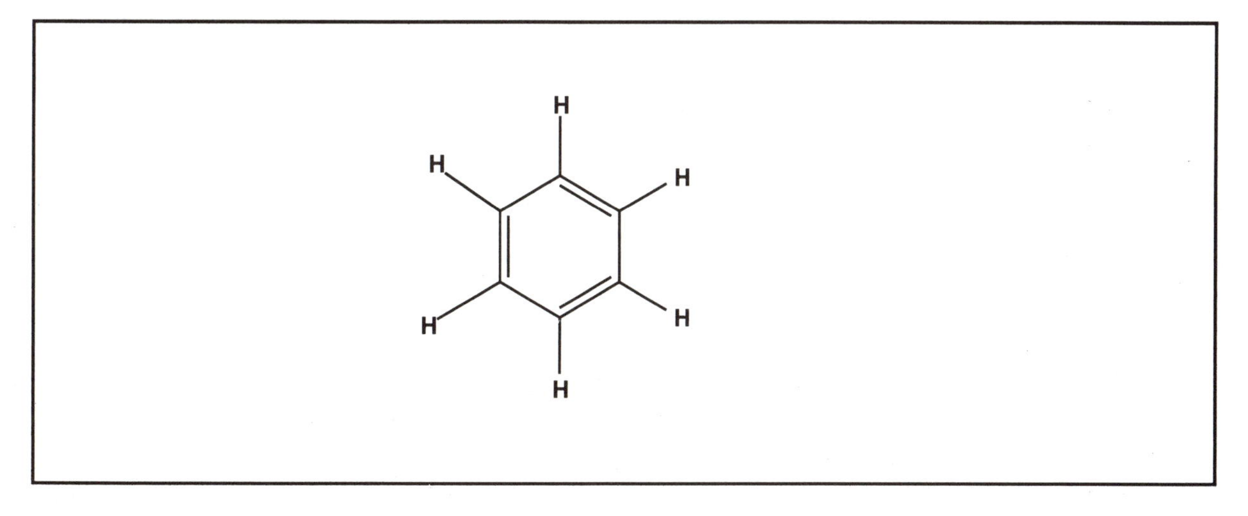

Answer

All the carbons are sp^2 hybridised. All the single bonds are σ bonds. Each double bond consists of one σ bond and one π bond.

Double bonds are shorter than single bonds. If benzene has the exact structure shown in the diagram, what kind of shape would the ring be?

Answer

The ring would be deformed with the single bonds longer than the double bonds.

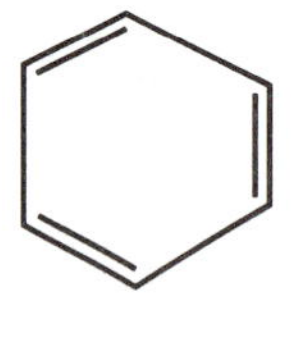

In fact the C–C bonds in benzene are all exactly the same length. In order to understand this, we need to look at the atomic and molecular orbitals. The following diagram shows benzene with all its σ bonds and is drawn such that you are looking into the plane of the benzene ring. Since all the carbons are sp^2 hybridised, there is a $2p_y$ orbital left over on each carbon. Draw the $2p_y$ orbitals onto the diagram and suggest how they could overlap with each other.

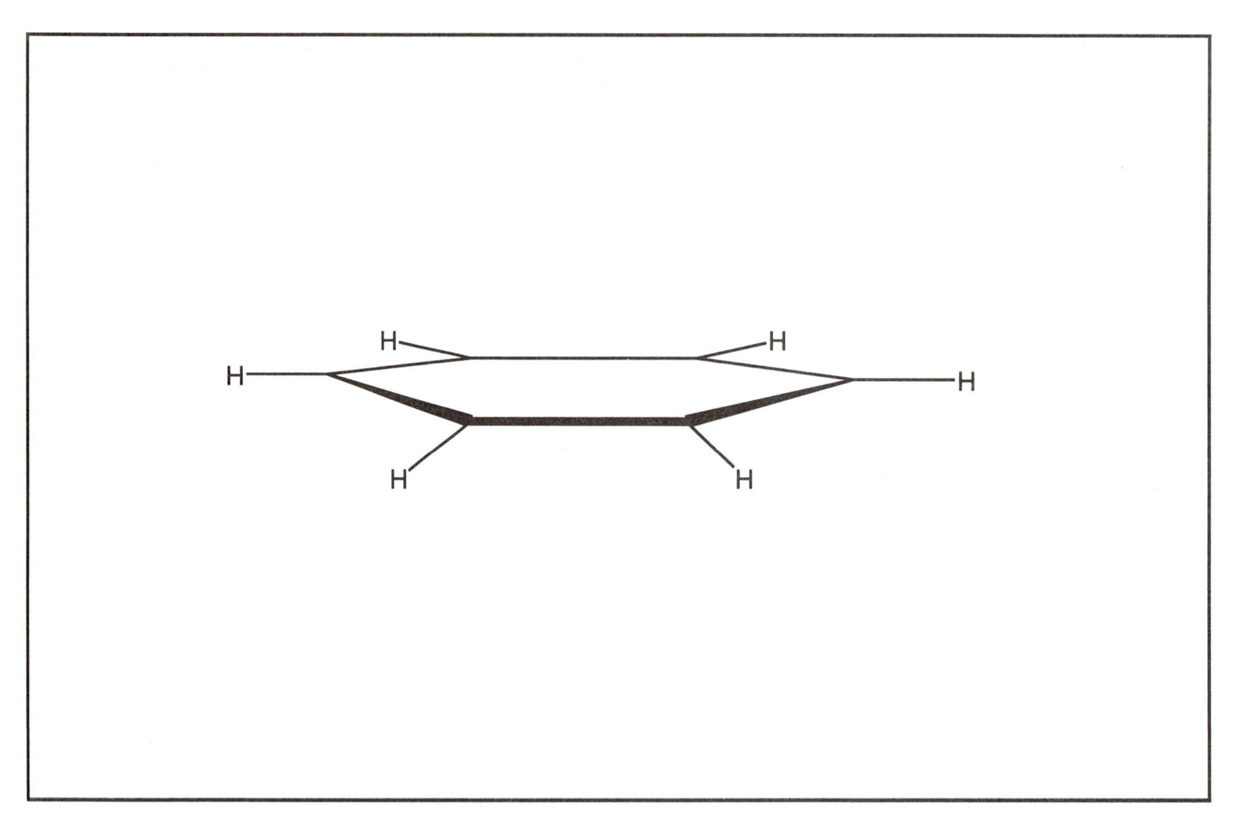

Answer

The $2p_y$ orbitals on each carbon can overlap with $2p_y$ orbitals on both sides of it.

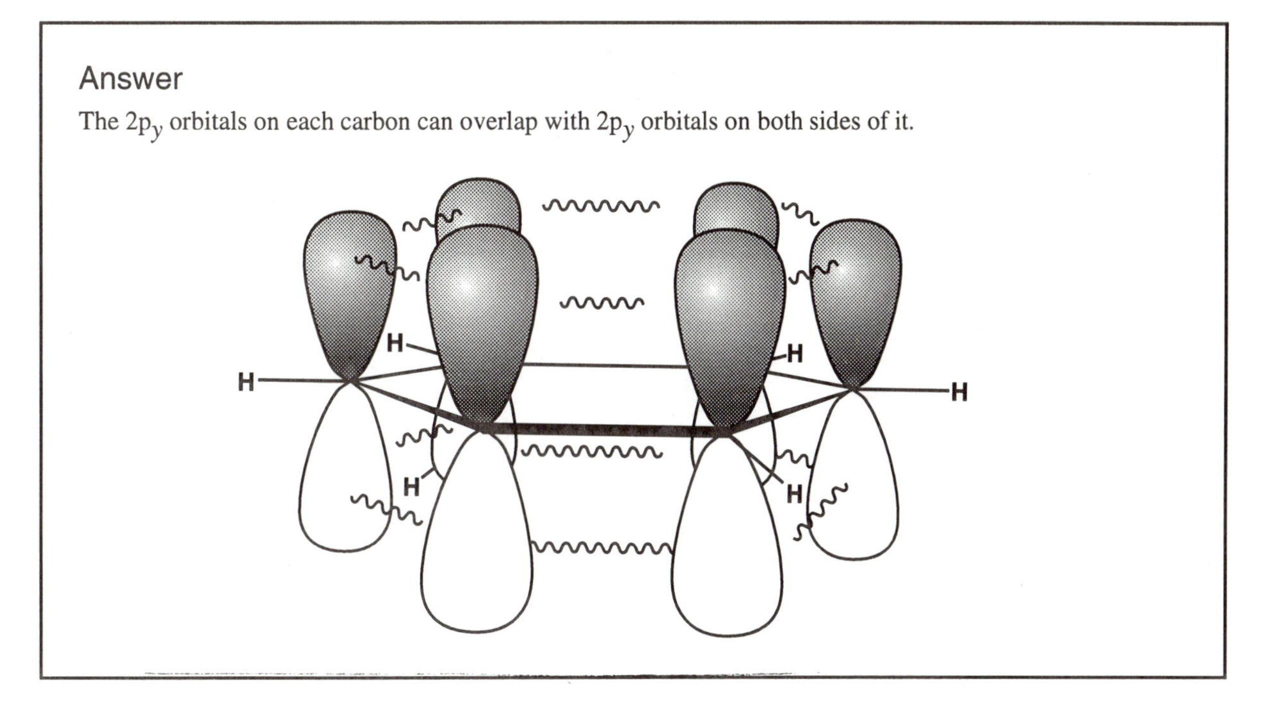

From this diagram, it is clear that the each $2p_y$ orbital can overlap with its neighbours right round the ring. This leads to a molecular orbital which involves all the $2p_y$ orbitals. Draw what shape you think this might be.

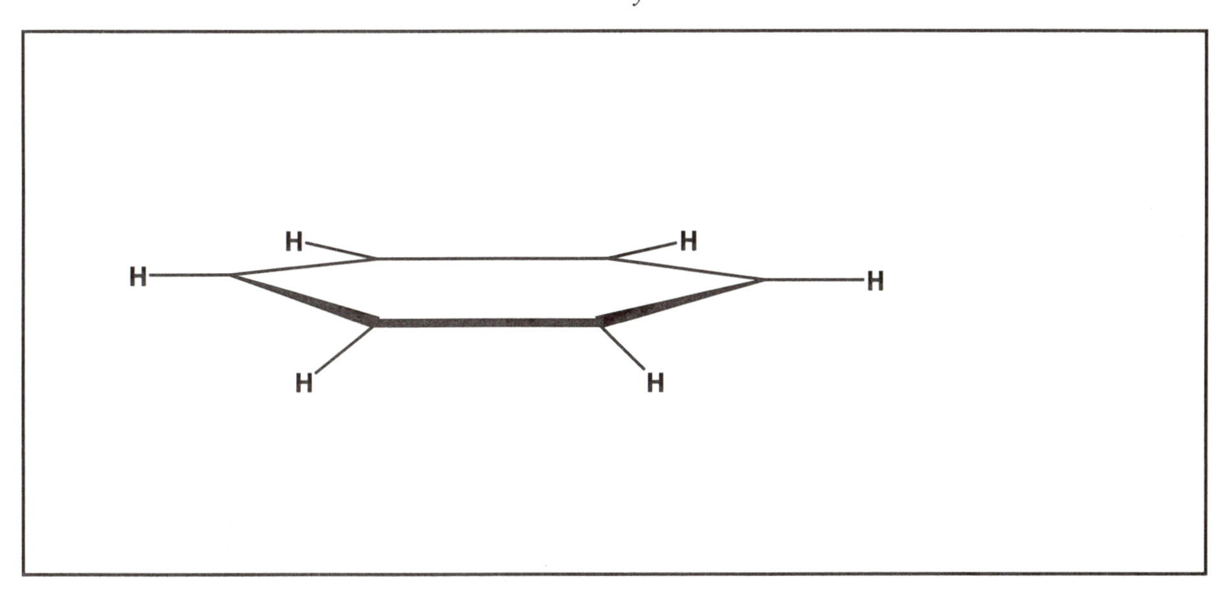

Answer

cont.

The upper and lower lobes of the $2p_y$ orbitals merge to give two doughnut like lobes above and below the plane of the ring. The molecular orbital is symmetrical and the six π electrons are said to be delocalised in the aromatic ring since they are not localised between any two particular carbon atoms.

The aromatic ring is often shown by the following diagram to represent this delocalisation of the pi electrons.

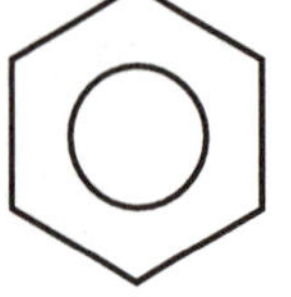

14.10 sp hybridisation

In sp hybridisation, the 2s orbital is mixed with one of the 2p orbitals (e.g. $2p_x$) which leaves two 2p orbitals unaffected ($2p_y$ and $2p_z$). How many sp hybrid orbitals are formed?

Answer

Two atomic orbitals were used in the hybridisation process. Therefore, there must be two hybrid sp orbitals resulting. The number of hybrid orbitals must equal the number of original atomic orbitals used for mixing.

Draw an energy diagram comparing the energies of the original atomic orbitals, the resulting sp hybridised orbitals and the remaining 2p orbitals. Place the electrons into their respective orbitals assuming that the energy difference between the sp orbitals and p orbitals is small.

Energy

Original atomic orbitals

sp hybridised orbitals

Answer

Since two atomic orbitals are used in the mixing or hybridisation process, there must be two sp hybrid orbitals of equal energy. The remaining 2p orbitals remain at their original energy level. In the diagram, these are the $2p_y$ and $2p_z$ orbitals.

cont.

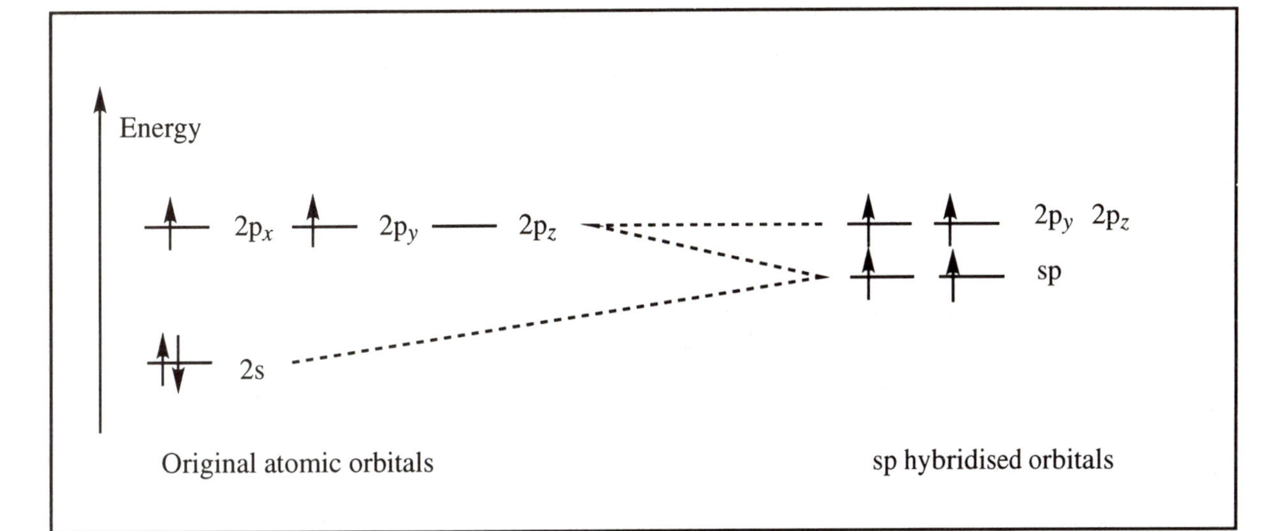

How many bonds are possible for an sp hybridised carbon. Would you expect them to be of equal energy or stability?

Answer

There are four half-filled orbitals and so you would expect four bonds. You would expect some difference in stability between the bonds with two strong σ bonds involving the sp orbitals and two weaker π bonds involving the $2p_y$ and $2p_z$ orbitals.

There are two sp^2 hybridised orbitals, a $2p_y$ orbital and a $2p_z$ orbital.
Draw the $2p_y$ and $2p_z$ orbitals in the following diagram.

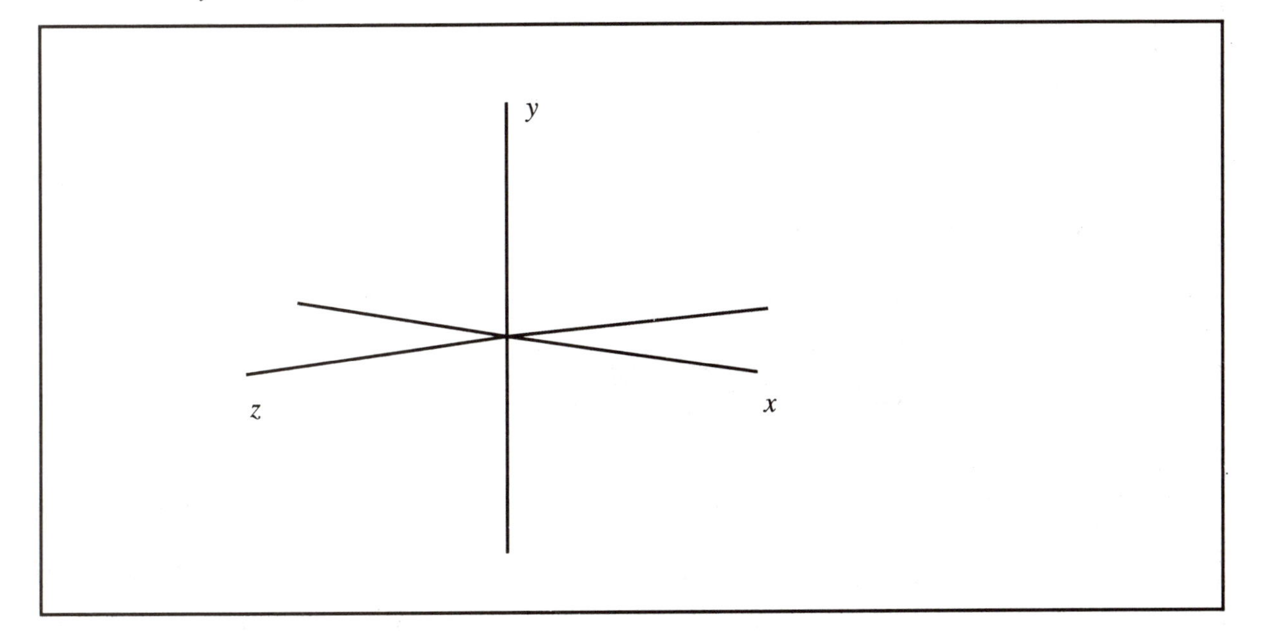

Answer

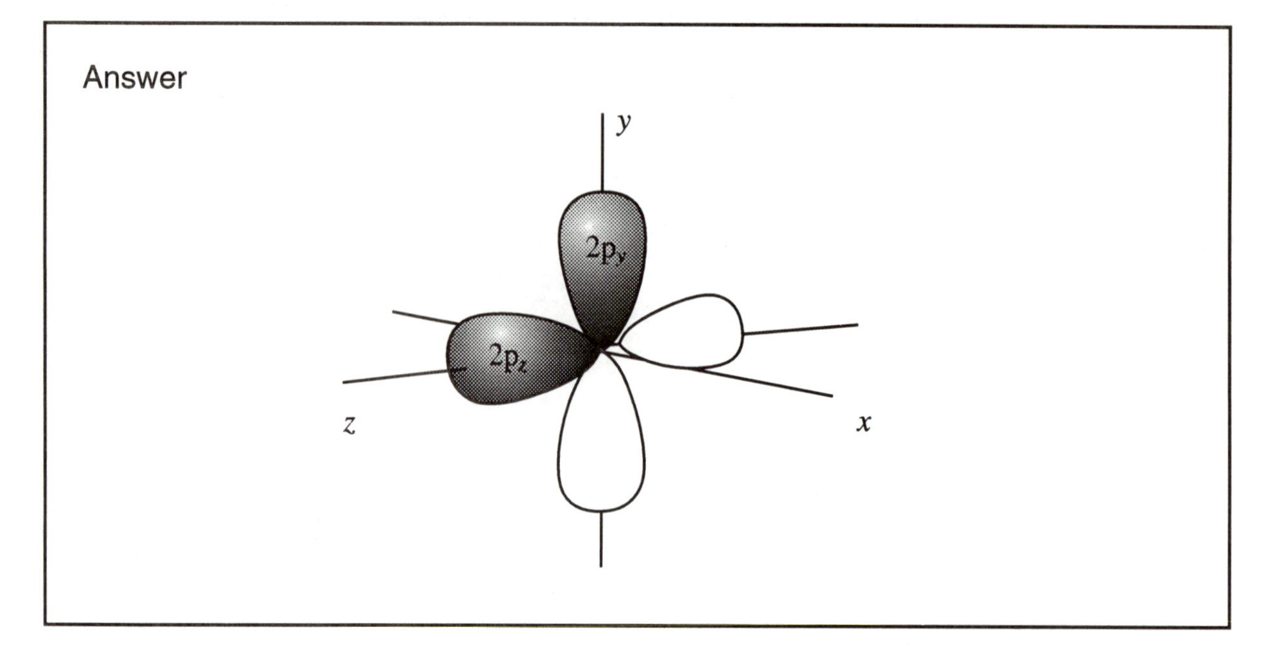

The space occupied by the 2p orbitals cannot be used by the sp orbitals. Taking this into account draw in the sp orbitals (major lobes only) such that they are as far apart from each other as possible.

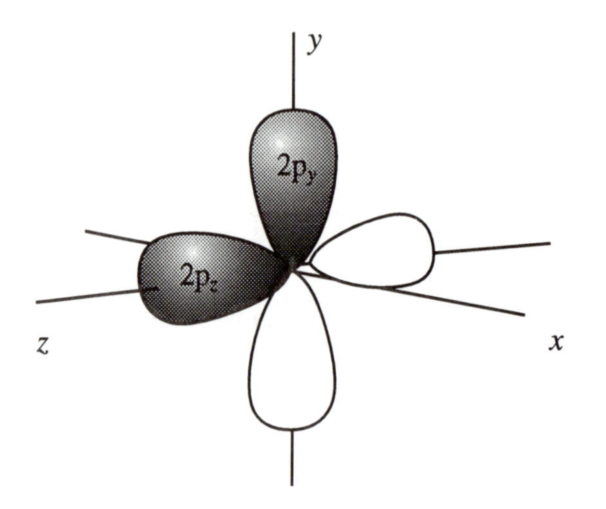

Answer

The two sp hybrid orbitals (in black) will both be in the *x* axis pointing in opposite directions.

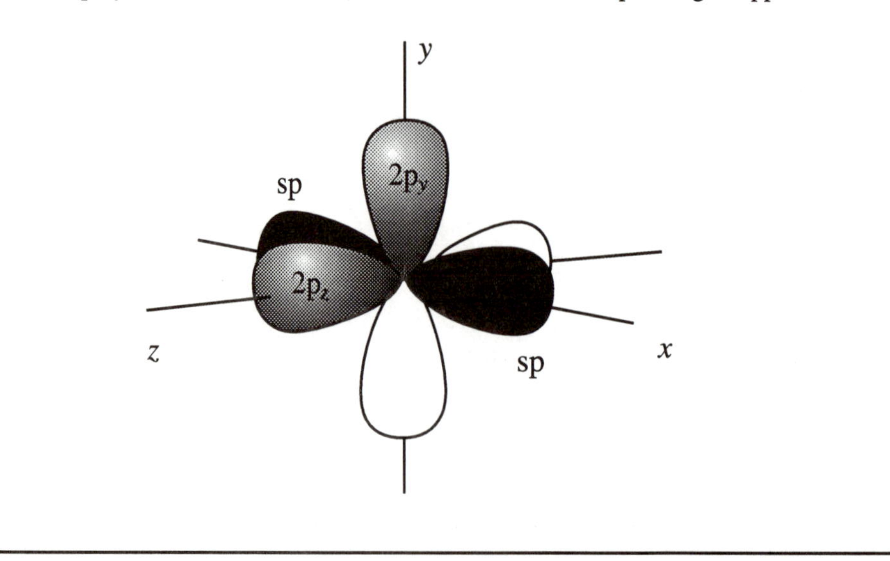

What shape of molecule would one get using the two sp orbitals for bonding?

Answer

A linear molecule.

We are now ready to look at how an sp hybridised carbon makes bonds in functional groups such as alkynes or nitriles.

14.11 Bonding in alkynes

Let us consider the bonding in ethyne where each carbon is sp hybridised. What orbitals are used in the formation of the C–H bonds? What kind of bond is formed?

Answer

The hydrogen atoms use their half-filled 1s orbitals to bond with a half-filled sp orbital on each carbon. The C–H bond is a σ bond which is a strong bond due to good overlap between the bonding orbitals.

If one of the sp orbitals on each carbon atom is used to form the C–H bond, what is the other sp orbital used for?

Answer

The remaining sp orbitals on both carbons is used to form a σ C–C bond.

The full σ bonding diagram for ethyne is now as shown.

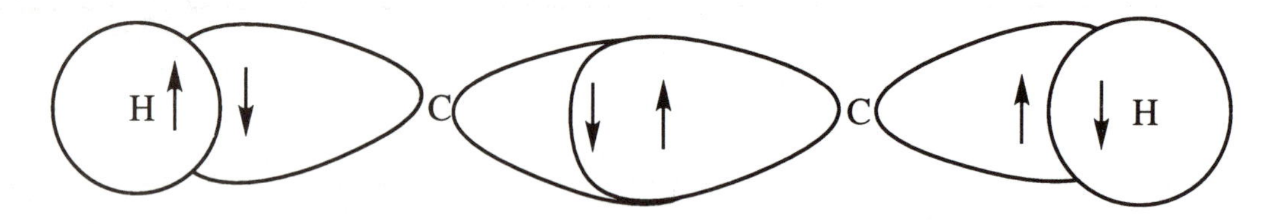

Each bond is a strong σ bond and we can simplify the diagram by drawing it as follows.

We know that ethyne is a linear molecule and we have seen how the σ bonds are made to give one of the triple bonds connecting the carbons in ethyne. We now use the remaining 2p orbitals to make the remaining two bonds. Draw in the $2p_y$ and $2p_z$ orbitals in the following diagram and show how they interact.

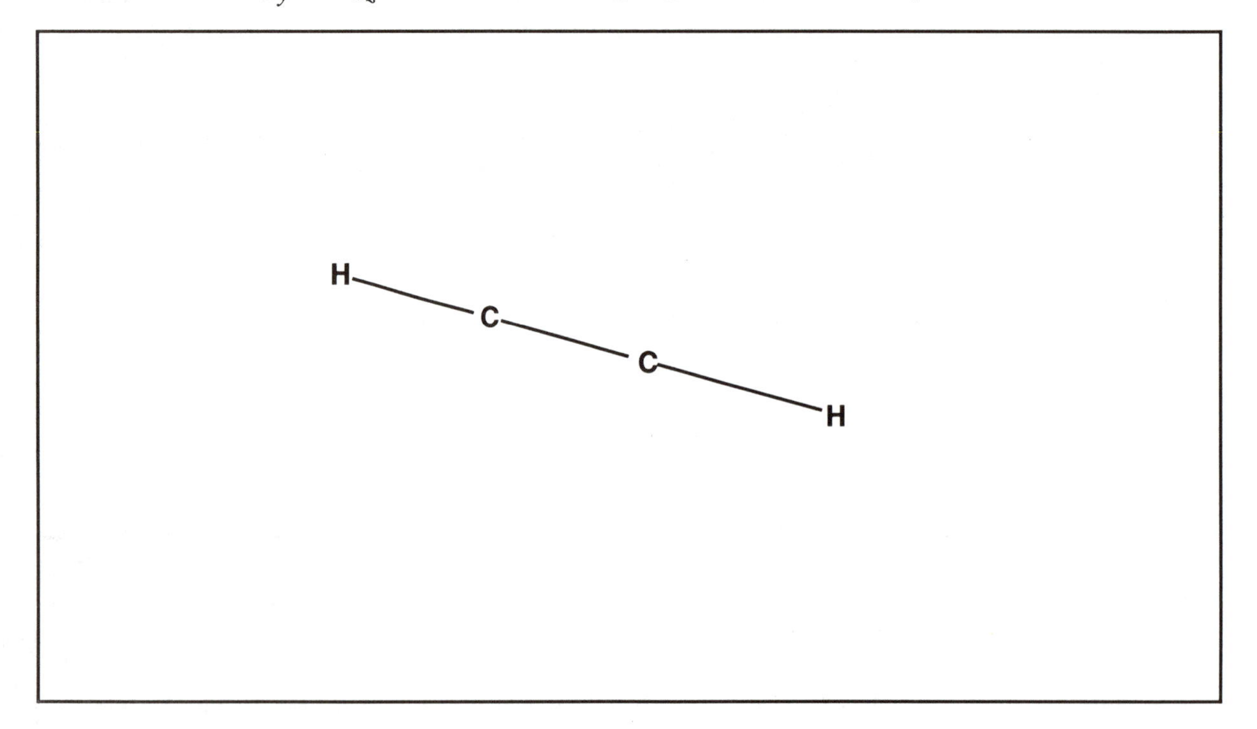

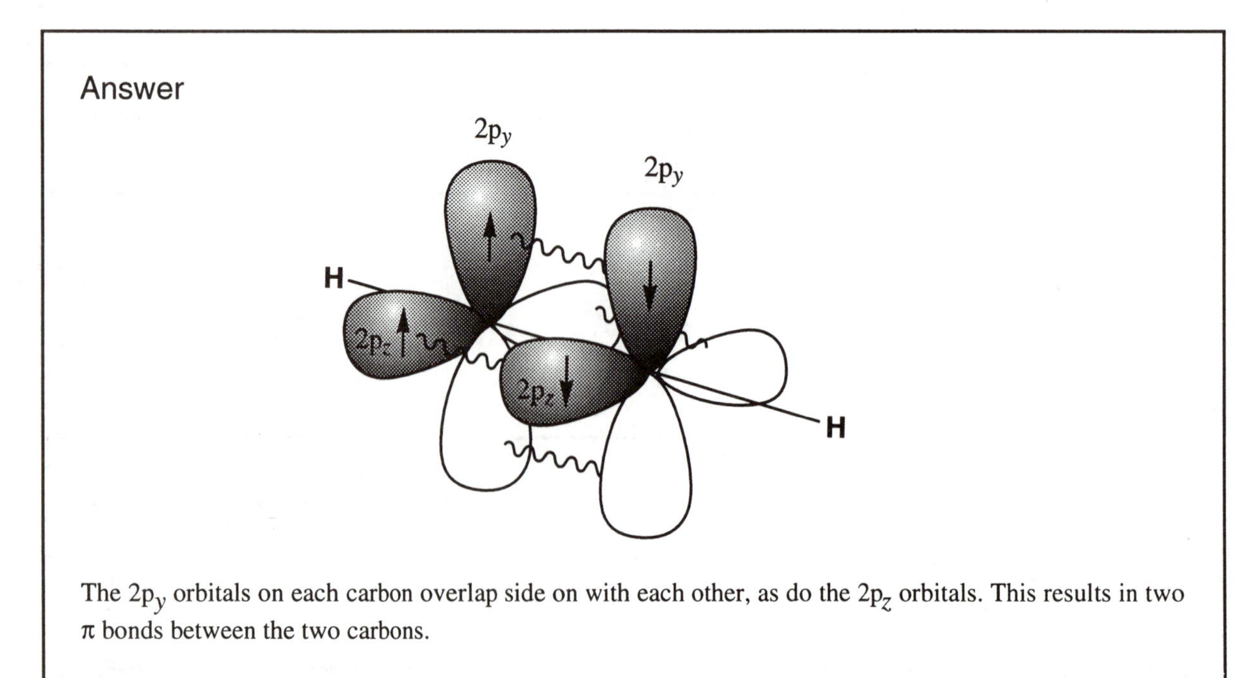

The $2p_y$ orbitals on each carbon overlap side on with each other, as do the $2p_z$ orbitals. This results in two π bonds between the two carbons.

Draw the two π orbitals on the following diagram.

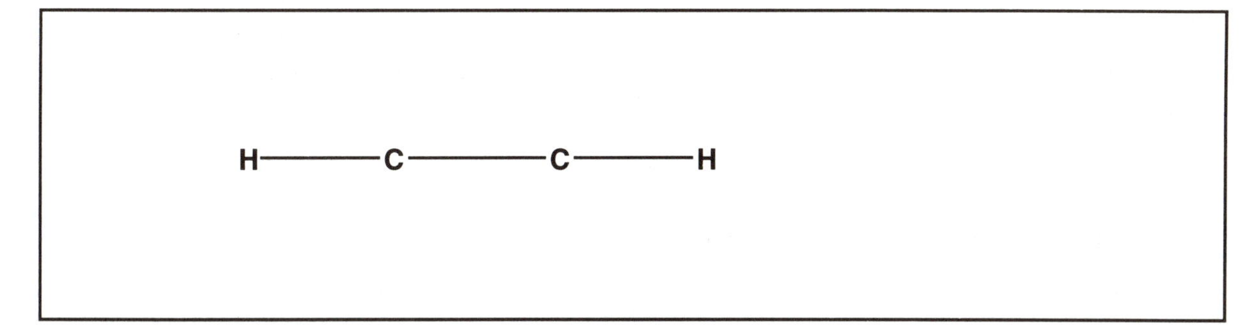

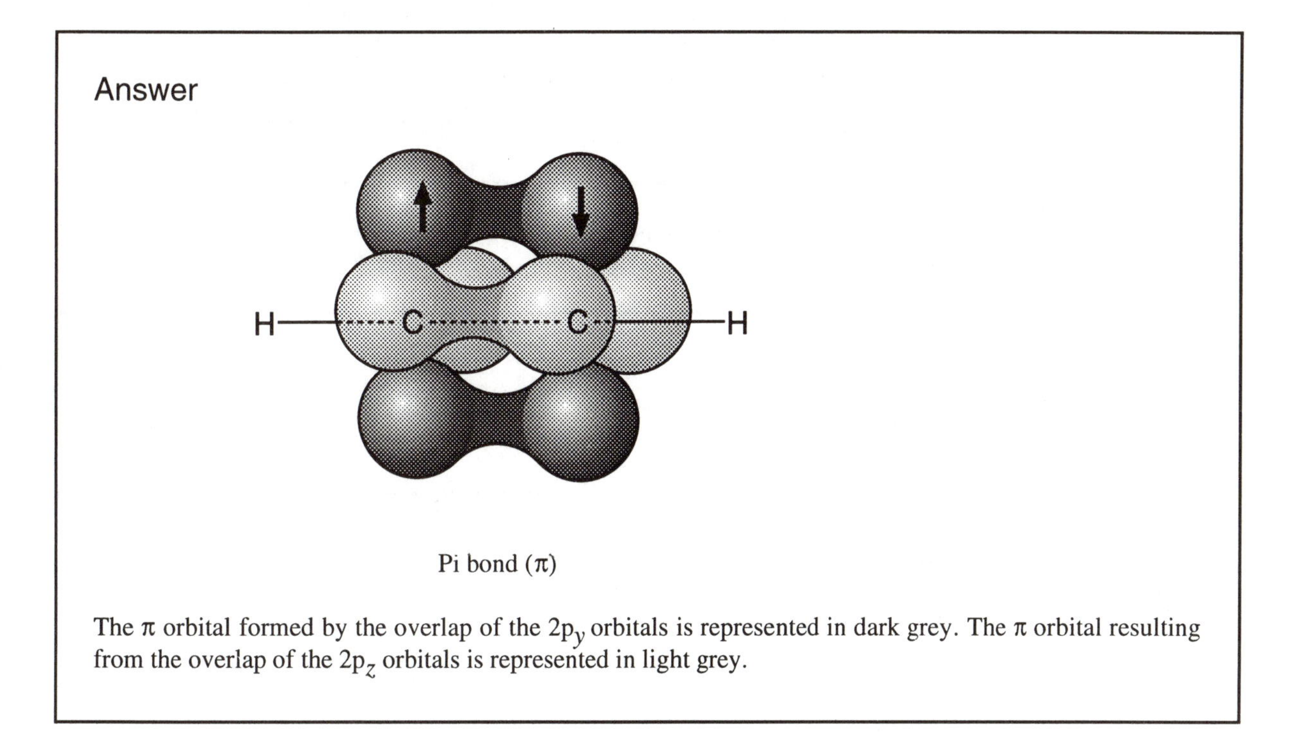

Pi bond (π)

The π orbital formed by the overlap of the $2p_y$ orbitals is represented in dark grey. The π orbital resulting from the overlap of the $2p_z$ orbitals is represented in light grey.

14.12 Bonding in the nitrile group (CN)

Exactly the same theory can be used to explain the bonding within a CN triple bond where both the carbon and the nitrogen are sp hybridised. Show in the following energy level diagram how the valence electrons of nitrogen are arranged after sp hybridisation.

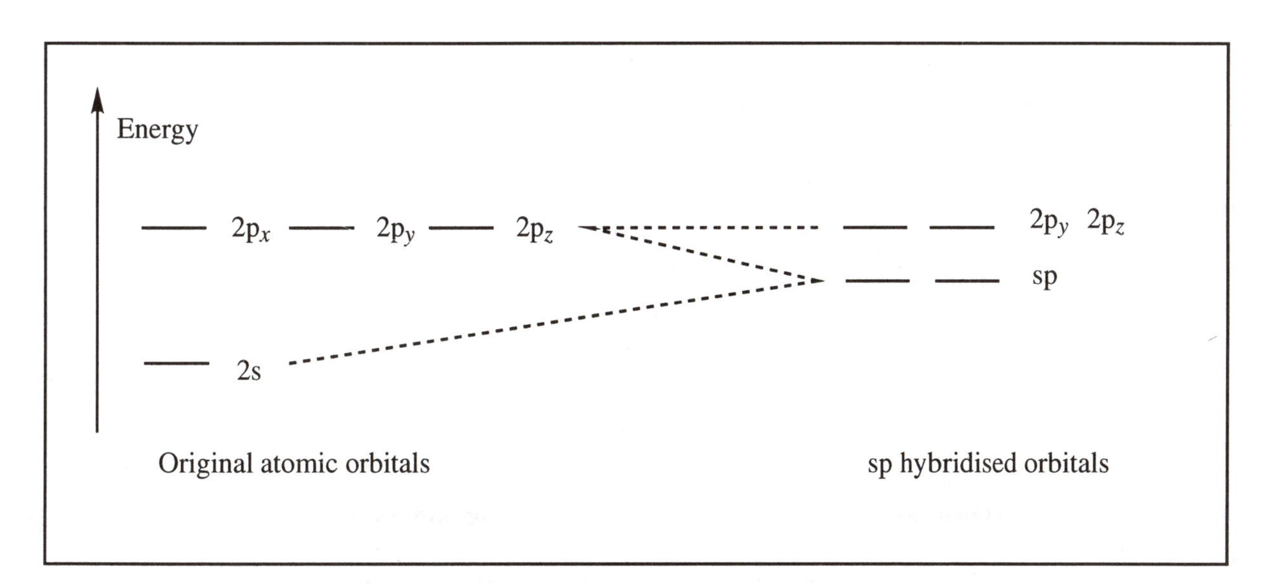

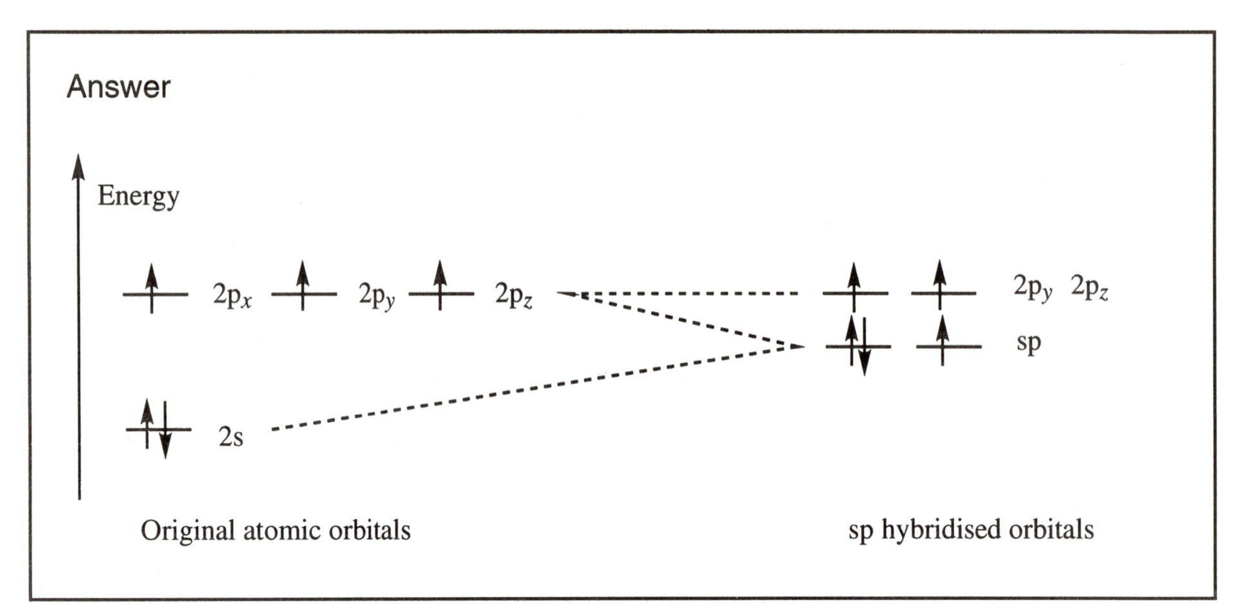

From this diagram, explain which orbital is used for the lone pair of electrons on nitrogen. Explain what kinds of bonds an sp hybridised nitrogen can form.

Answer

The lone pair of electrons is in one of the sp orbitals. The other sp orbital can be used for a strong σ bond. The $2p_y$ and $2p_z$ orbitals can be used for two π bonds.

The following diagram shows the σ bonds of HCN. Show how the remaining 2p orbitals are used to form two π bonds.

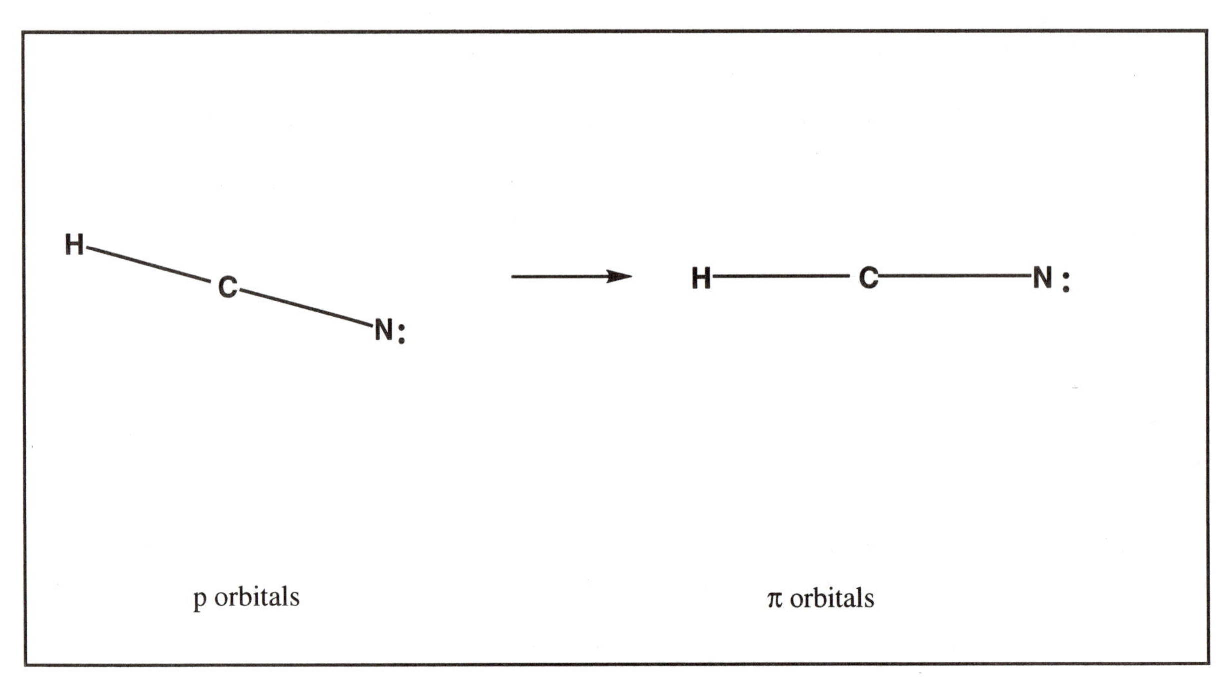

Answer

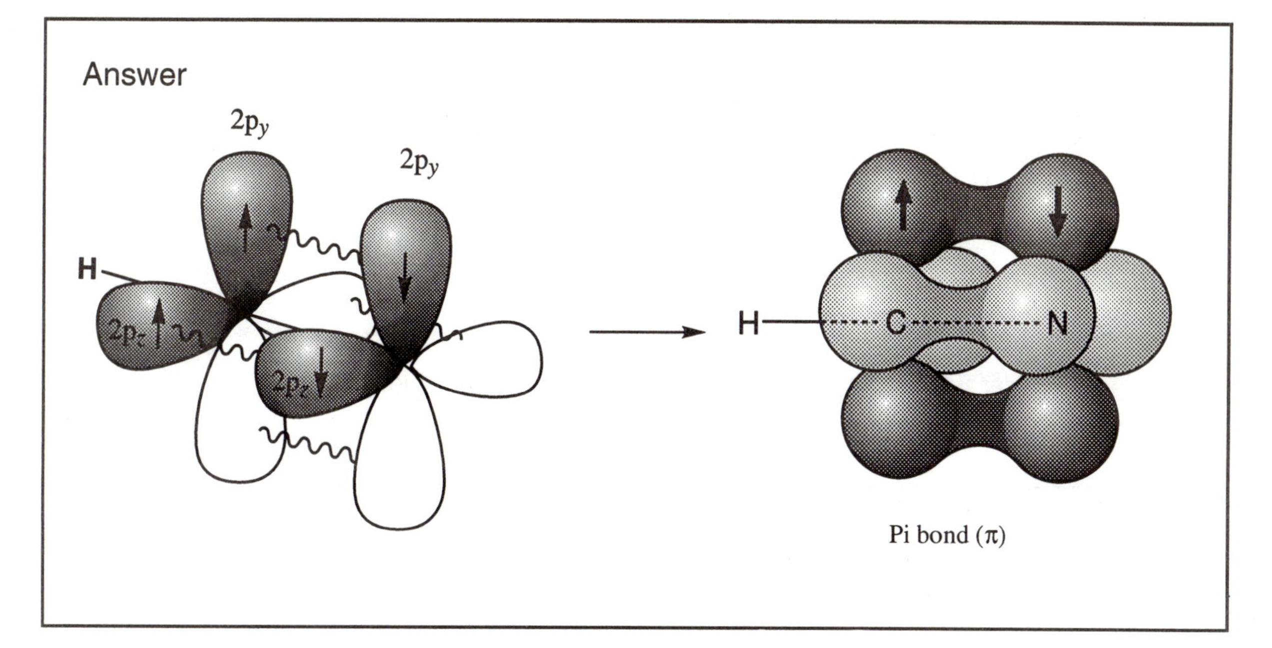

Pi bond (π)

14.13 σ versus π bonding

A favourite exam question is to ask you to identify σ and π bonds in a molecule. This is quite easy as long as you remember a few rules.

- All bonds in organic structures are either σ or π bonds.
- All single bonds are σ bonds resulting from strong direct overlap of orbitals.
- All double bonds are made up of one σ bond and one π bond, the latter resulting from the side on overlap of p orbitals.
- All triple bonds are made up of one σ bond and two π bonds.

Identify the sigma and pi bonds in the following structures. (Identify only those bonds shown.)

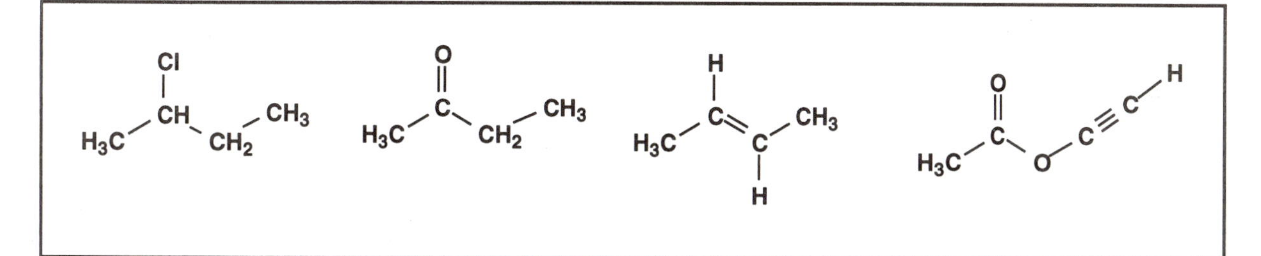

Answer

All the bonds shown are σ bonds except those labelled as π bonds.

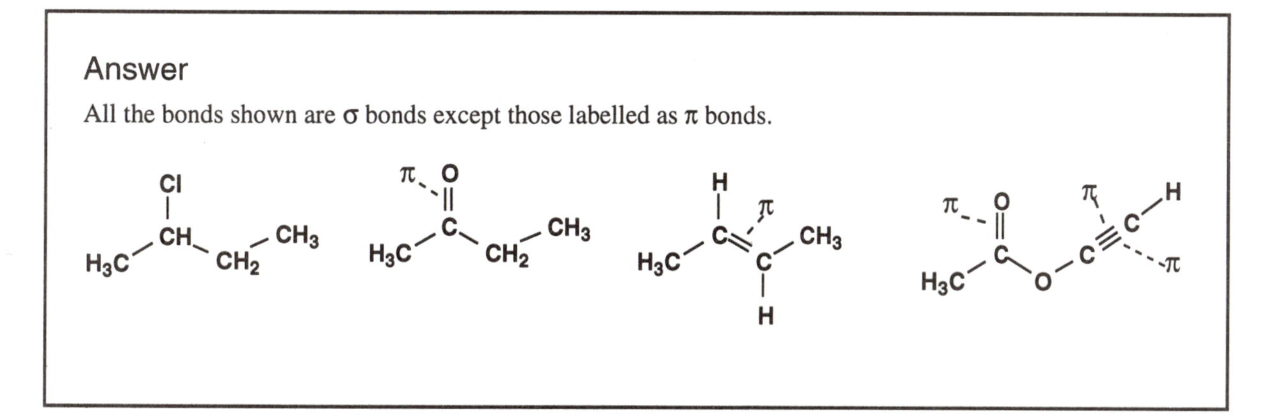

14.14 Hybridised centres (sp, sp^2 and sp^3)

Another favourite exam question is to ask you to identify sp, sp^2 and sp^3 centres. Remember the following rules.

- Both carbon atoms involved in the double bond of an alkene (C=C) must be sp^2 hybridised.
- Both the carbon and the oxygen of a carbonyl group (C=O) must be sp^2 hybridised.
- All aromatic carbons must be sp^2 hybridised.
- Both atoms involved in a triple bond must be sp hybridised.
- All atoms involved in single bonds only must be sp^3 hybridised (except hydrogen).
- Hydrogen uses a 1s orbital for bonding and is not hybridised.

Identify the sp, sp^2 and sp^3 centres in the following molecules.

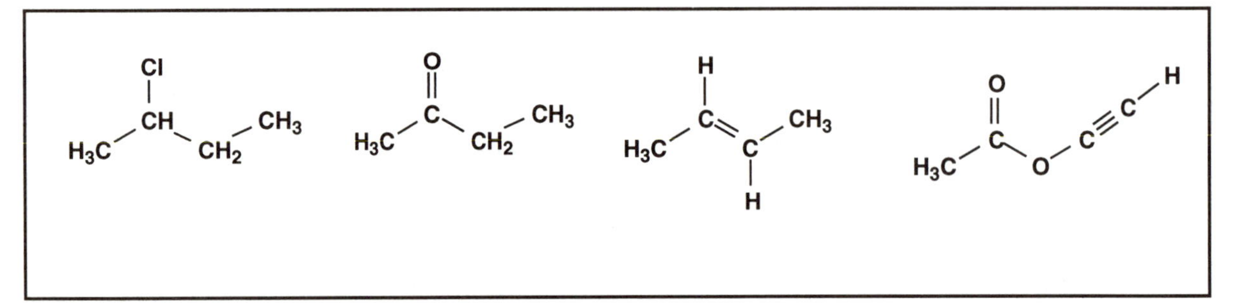

Answer

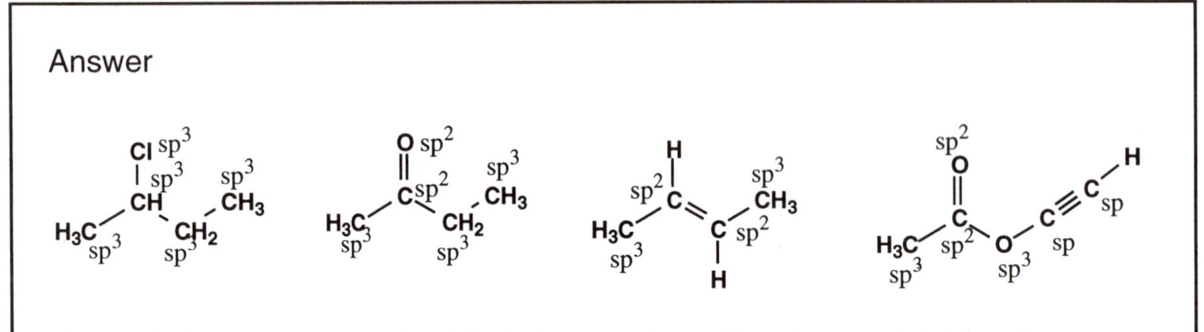

Note that hydrogen atoms are not hybridised. They can only bond by using an s orbital since there are no p orbitals in the first electron shell. It is therefore impossible for a hydrogen to take part in π bonding.

Oxygen and halides on the other hand can form hybridised orbitals which are either involved in bonds or in holding lone pairs of electrons.

14.15 Shape

All sp^3 hybridised centres are tetrahedral. All sp^2 hybridised centres are trigonal planar. All sp centres are linear. This results from the space required by the π bonds.

The shape of functional groups is determined by the bonding. Functional groups containing sp^2 centres will be planar. Note that it is the functional group which is planar and not the whole molecule. Identify the functional groups you have met which are planar and those which are linear.

Planar functional groups:

Linear functional groups:

Answer

Planar functional groups:

aldehyde, ketone, alkene, carboxylic acid, acid chloride, acid anhydride, ester, amide.

Linear functional groups:

alkyne, nitrile.

The following molecules contain planar functional groups. Identify the atoms which are in the plane. (It helps to use molecular models for an exercise like this.)

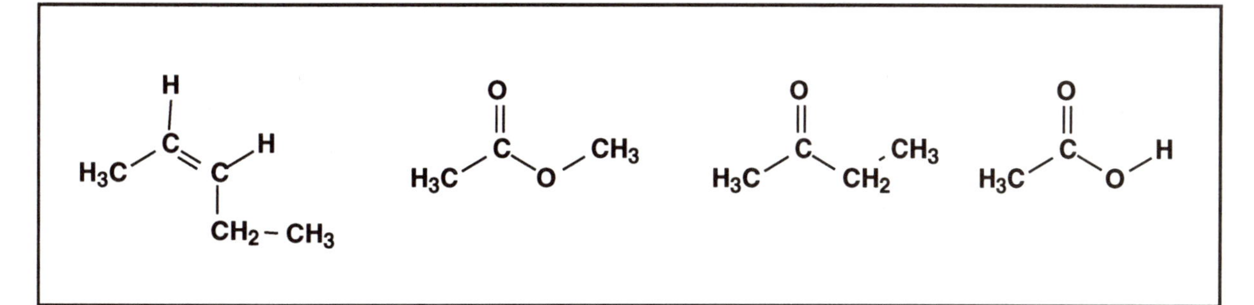

Answer

Only the atoms circled are in the plane of the functional group.

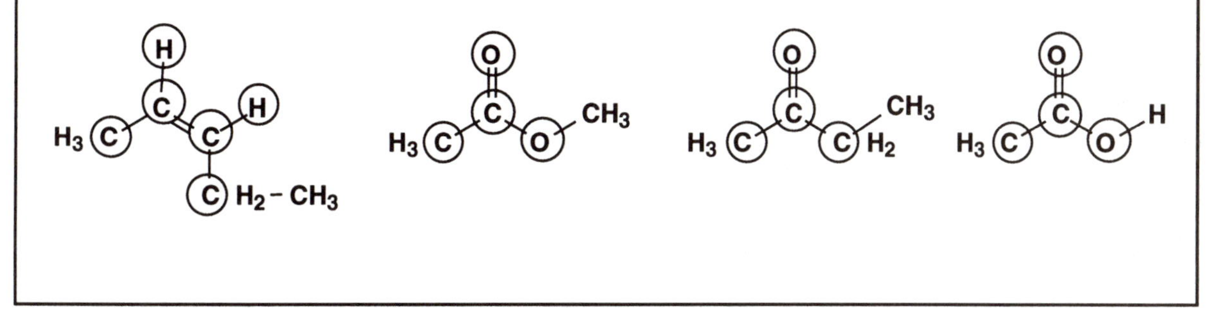

Summary

- Single bonds in organic molecules are called sigma bonds and are formed from the direct overlap of sp, sp^2 or sp^3 hybridised orbitals with another hybridised orbital or with the 1s orbital of hydrogen.
- Double bonds are made up of one σ bond and one π bond.
- Triple bonds are made up of one σ bond and two π bonds.
- Pi bonds are formed from the side on overlap of p orbitals which have not been hybridised.
- An sp^3 centre is an atom where the s orbital has been hybridised with all three p orbitals to give four identical sp^3 orbitals. An sp^3 hybridised carbon will have four σ bonds.
- An sp^2 centre is an atom where the s orbital has been hybridised with two of the three p orbitals to give three identical sp^2 orbitals. An sp^2 hybridised carbon will have three σ bonds. The remaining p orbital will be used to form a π bond to one of the atoms already linked by a σ bond. Therefore, an sp^2 hybridised carbon will have two single bonds and one double bond.

- An sp centre is an atom where the s orbital has been hybridised with one of the three p orbitals to give two identical sp orbitals. An sp hybridised carbon will have two σ bonds. The remaining p orbitals will be used to form two π bonds to one of the atoms already linked by a σ bond. Therefore, an sp hybridised carbon will have one single bond and one triple bond.†
- Functional groups which have single bonds and sp^3 hybridised carbons will be tetrahedral at the carbon (alcohols, ethers, alkyl halides).
- Functional groups which have double bonds and sp^2 hybridised carbons will be planar (alkenes, aldehydes, ketones, carboxylic acids and carboxylic acid derivatives).
- Functional groups which have triple bonds and sp hybridised carbons will be linear (alkynes, nitriles).

† Functional groups known as allenes ($R_2C{=}C{=}CR_2$) have an sp hybridised carbon located at the centre of two double bonds, but these functional groups are beyond the scope of this text.